Union Public Library

S0-ANV-898

MODERN WORLD NATIONS

AFGHANISTAN	INDIA
AUSTRIA	IRAN
BAHRAIN	IRAQ
BERMUDA	IRELAND
BRAZIL	ISRAEL
CANADA	JAPAN
CHINA	KAZAKHSTAN
COSTA RICA	KUWAIT
CROATIA	MEXICO
CUBA	NEW ZEALAND
EGYPT	NORTH KOREA
ENGLAND	PAKISTAN
ETHIOPIA	RUSSIA
REPUBLIC OF GEORGIA	SAUDI ARABIA
GERMANY	SCOTLAND
GHANA	SOUTH KOREA
ICELAND	UKRAINE

Ireland

Edward P. Hogan
South Dakota State University

and

Erin Hogan Fouberg
Mary Washington College

Series Consulting Editor
Charles F. Gritzner
South Dakota State University

'nion Public Library

CHELSEA HOUSE
P U B L I S H E R S
A Haights Cross Communications Company
Philadelphia

Frontispiece: Flag of Ireland

Cover: Sheep grazing near the coast in County Clare, Ireland.

CHELSEA HOUSE PUBLISHERS

VP, New Product Development Sally Cheney
Director of Production Kim Shinners
Creative Manager Takeshi Takahashi
Manufacturing Manager Diann Grasse

Staff for IRELAND

Executive Editor Lee Marcott
Production Editor Jaimie Winkler
Picture Researcher 21st Century Publishing and Communications, Inc.
Cover and Series Designer Takeshi Takahashi
Layout 21st Century Publishing and Communications, Inc.

©2003 by Chelsea House Publishers, a subsidiary of Haights Cross Communications.
All rights reserved. Printed and bound in the United States of America.

A Haights Cross Communications Company

http://www.chelseahouse.com

First Printing

1 3 5 7 9 8 6 4 2

Library of Congress Cataloging-in-Publication Data

Fouberg, Erin Hogan.
  Ireland / by Erin Hogan Fouberg and Edward Patrick Hogan.
     p. cm.—(Modern world nations)
Includes index.
  ISBN 0-7910-7377-7
  1. Ireland—Juvenile literature. [1. Ireland.] I. Hogan, Edward Patrick, 1939–
II. Title. III. Series.
DA906 .F68 2002
941.7—dc21
                                                                              2002015902

Table of Contents

Ireland

Crop fields, pastures, and fences on the Fanad Peninsula in County Donegal typify Ireland's rural heritage. County Donegal is located on Ireland's northwest coast. The Atlantic Ocean's consistent temperature helps maintain Ireland's moderate climate.

Introduction

Triall ó Dhealbhna
Land of warriors and of bards,
Land of clear blue streams with birds,
Banba of girls with golden tresses,
And of brave young men.

By Ulilliam Nuiseann
Published by: Triona Pers, Netherlands

Perception is a key to understanding the geography of Ireland. Images or perceptions can be created from pictures and stories about Ireland or from family, television, books, and newspapers.

For many people, Ireland is perceived as a land of eloquent poets, renowned writers and playwrights, lyrical musicians, and creative dancers. Among these many talented literary and artistic geniuses are poets Oscar Wilde and William Butler Yeats, writers Padraic Pearse and James Joyce, musicians the Chieftains and U2, and Irish dancers such as those in *Riverdance.*

If students and teachers in your school were to be asked for their perceptions when they hear the word "Ireland," answers may include: the Emerald Isle, forty shades of green, leprechauns, potatoes, the Great Famine, stone fences, very narrow roads, where grandpa was born, or The Troubles (civil strife, primarily in Northern Ireland). The quote at the beginning of this chapter reflects many of the common perceptions people have of Ireland, its geography, and its history.

Geographers study places, and it is clear that the way a place is perceived will change when that place is experienced. If students or teachers in your school who have recently visited Ireland were asked what they think of when they hear the word "Ireland," they would have completely different perceptions. They would talk about motor coaches everywhere, new highways, big new homes in every community, the Irish language, lots of jobs, new factories across the countryside, the euro (currency of the European Union), software companies, golf courses, and thousands and thousands of tourists.

Today's Ireland is a land of rapid change. The purpose of this book is to help you understand the rapid changes that are occurring in Ireland and to consider the future impact on the land and the people. Places change over time in numerous ways. When an economy changes from agricultural to high technology, it changes the landscape—new buildings go up, and farms turn into suburban subdivisions.

Understanding the numerous, and sometimes conflicting, perceptions of Ireland requires critical thought. Geographic theories and concepts will be used as the bases of this analysis of Ireland. While reading the book, try to ask why things are located where they are in Ireland both historically and today.

To become a good geographer, one needs to know what geographers do and how the world and individual countries within it, such as Ireland, are studied. The basic belief of geography is that all features (whether physical or cultural) are organized rationally on Earth's surface. Geography explains why hills, rivers, and even gale-force winds are where they are. Geography also explains why Celtic sacred sites, the Catholic religion, and Irish dancers are where they are. The physical processes of erosion, such as wind and water, help to explain why a mountain or river exists. The human and cultural processes of change, such as diffusion and migration, help explain why a language is lost or reborn.

Geography is an exciting field because the earth is constantly changing. Over centuries, people build towns, change the courses of rivers, and create new economic systems. These human changes reflect each people's culture—their way of life—including their customs, values, material objects, and their unique place in the world. At the same time, changing physical processes such as mountain building, erosion, and deposition affect humans and their cultures. Humans and environments interact in a reciprocal relationship. Humans must adapt to environments, they also depend upon natural resources within the environment and, in countless ways, people change the environment.

LOCATION

The island of Ireland (Eire) is located at the western limit of the continent of Europe. The Irish Sea separates Ireland from the British Isles to the east and northeast. The Atlantic Ocean surrounds Ireland on the northwest, west, and south. Because of its rugged coastline, no place in Ireland is more than 70 miles (113 kilometers) from the sea.

From north to south, Ireland stretches 302 miles (486 kilometers), located roughly between 52 and 56 degrees north latitude (approximately equivalent to Canada's Lake Winnipeg).

From east to west, Ireland reaches 171 miles (275 kilometers), between approximately 6 and 11 degrees west longitude. The island has an area of 32,589 square miles (84,405 square kilometers), about the size of Maine. However, the island of Ireland is divided into two political units: the Republic of Ireland (Ireland) and Northern Ireland. Ireland is an independent country of 27,136 square miles (70,282 square kilometers), slightly larger than West Virginia. Northern Ireland is an Administrative Division of the United Kingdom and covers 5,453 square miles (14,123 square kilometers), slightly larger than Connecticut.

OVERVIEW

Although Ireland is an island off the coast of Europe, its history is tied closely to the European region. Long occupied by the Celtic people who ranged from Ireland to Italy, Ireland still traces much of its heritage to its Celtic roots.

Over the last 2,000 years, Ireland has experienced what many of the poorest countries in the world have been through in the last two hundred years. Although Ireland was never part of the Roman Empire, it was influenced by the Romans, who brought Catholicism. Ireland was later colonized by the Vikings and then by the British. More recently, Ireland suffered the great potato famine, resulting in at least two million migrants leaving Ireland and relocating in North America, Australia, and New Zealand. With that great migration went many Irish speakers, and then Ireland's language started to decline.

The effect of immigration from Ireland is seen in the United States today, where over 50 million people (about 22 percent of the population) claim to be of Irish descent. The island of Ireland still experiences the lingering effects of colonialism in its division into two political units: the Republic of Ireland (an independent country since 1921) and Northern Ireland (still part of the United Kingdom). The political instability in Northern Ireland is a constant reminder to the

Located on the western limit of the continent of Eurasia, the island of
Ireland (Eire) consists of the Republic of Ireland and, in the northeast,
Northern Ireland which is an administrative unit of the United Kingdom.
About one person in four of Ireland's nearly 4 million inhabitants lives
in or near the Irish capital of Dublin.

Irish of the lasting changes that come from colonization.

The poorest countries of the world struggle to find a way to build businesses that will bring jobs, sustainable development, and stable governments. Even following political independence, the economy of former colonies is tied to the colonizer and other wealthy countries. No matter what a country produces, the biggest consumers of its products live in the wealthiest countries of the world, because they have the money to purchase goods and resources. Today, the wealthiest 20 percent of the world controls 83 percent of the world's wealth.

In the last 25 years, Ireland has gone from one of the poorest countries in Western Europe to one of the world's fastest-growing economies. The roles the European Union, the Irish government, and foreign investors have played in creating this economic miracle will be examined in this book. Such a rapid change in a country's economy is uncommon, and why it has happened at this time in Ireland will also be analyzed.

The story of Ireland's economic change is just one of the many stories in this book. A unique part of Irish culture is the numerous tales, myths, legends, and folklore passed down orally through storytelling. Perhaps the most famous are the stories of Cuchulainn the warrior and the children of Lir who were turned into swans by their mother. Hundreds of other Irish legends, such as the earth-shapers, Saint Brigid's cloak, and MacDatho's boar, can fill one's mind with perceptions of Ireland. These fascinating legends, however, are not the focus of this book.

Each chapter provides a different story about Ireland. Each story will also change many perceptions of that place. The massive changes that occurred in Ireland over the last century are reflected in its physical and cultural landscapes. Ireland has a fascinating physical geography. Its cultural geography is a microcosm reflecting many of the changes that are happening today throughout much of the world. This book's goal is to use stories about how Ireland has changed over time and to present a new story that describes the Ireland of today and tomorrow. When countries change, the people often struggle over whether to remain traditional or to fully embrace the new. At the end of this book, some perceptions of Ireland will remain, and some valuable new perceptions will have developed.

As the story of Ireland's geography commences, the lead of the great Irish storytellers will be followed, beginning with the Irish words, "*fadó . . .* fadó . . . ," *meaning, "long, long ago."*

Situated in County Clare on Ireland's western coast, the Cliffs of Moher stretch for five miles (eight kilometers) and reach a height exceeding 700 feet (214 meters). The Cliffs of Moher are composed of several types of sedimentary rock; the hardest, limestone, comprises the protective cap rock that helps the cliffs resist erosion from the wind and ocean waves.

2

Ireland's Physical Environment

Fadó, fadó, Ireland's geologic roots began with the formation of Western Europe. The island of Ireland today is a direct result of the earth-building forces that created Eurasia over two billion years ago. Over time, layers of sediment accumulated on the floor of ancient geologic seas, creating limestone. A mixture of this limestone and lava settling on the ocean floor formed the island of Ireland. At the same time these earth-building forces constructed the island, the forces of erosion have worked to wear it down.

Ireland is not one homogeneous physical area, but rather a complex land. It has experienced mountain building through faults and folds, uplifts, and magma flows. All the landscapes resulting from earth building are subject to erosion by the forces of ice, water, and wind. Streams and waves erode materials from one place and deposit them in another. Ice Age glacial flows bulldozed and scoured the surface as they expanded and dumped debris as they retreated.

Today, Ireland is a land of plains and highlands. The plains are in the center of the island, and the hills are in the coastal zones. The best way to envision the physical geography of Ireland is to think of it as resembling a cereal bowl. Ireland has a low, broad interior, the bottom of the bowl, and is surrounded by highlands, the rim of the bowl.

TERRAIN

The broad interior is called the central lowlands. It contains extensive glacial debris, peat bogs, rivers, lakes, and coastal shores. It is Ireland's most productive agricultural zone. Dublin, the capital and largest city in Ireland, is located where the central lowlands region reaches the east coast.

On the rim of this lowland are hills, low mountains, and steep sea cliffs. These highland areas are more difficult to navigate. They include small villages and towns and typically less productive soils.

Overall, processes of physical geography in Ireland have resulted in a small island connected to the geologic history of Western Europe. At the end of the Ice Age (Pleistocene Period) over 10,000 years ago, the glaciers receded from Europe, depositing sediment, called glacial drift, and forming lakes, rivers, and fjords. While there are few fjords (such as the Killary fjord between counties Mayo and Galway in Ireland), many of the lakes and rivers of the central lowlands were formed at the end of the Ice Age.

The mountains of northwest Ireland, while covered with glacial drift, are actually part of a mountain range that stretches from Scandinavia to Scotland and Ireland. This mountain chain is the eroded remnants of the ancient Caledonian mountains, covered in places by the Atlantic Ocean and Irish Sea. In Ireland, the mountains extend through the island's northwest. Along the coast, their rocky cliffs are still subject to the scouring forces of erosion.

The mountains and highlands of southern Ireland are remnants of the Armorican mountains that once extended from Ireland eastward to central Europe. The Armorican highland landscape varies in Ireland from west to east. In the west, in County Kerry and the western part of County Cork, the landscape contains high sandstone mountains, numerous peninsulas, and flooded lowlands and inlets. The Macgillycuddy Reeks Range in this region contains Carrauntuohil, which at 3,414 feet (1,041 meters) is the highest mountain in all of Ireland.

The Armorican mountains in eastern Cork and Waterford counties are composed of sandstone, limestone, and shale. Through water erosion, ridges and valleys have formed parallel to each other, extending from west to east. The ridges are comprised of resistant sandstone rock. Water eroded the limestone and shale, creating the valleys. As a result, the ridges are hill-like, and streams occupy the lower valleys. The streams in this area have developed a definite trellis drainage pattern. The trellis pattern forms when stream erosion forms water gaps by cutting through less resistant rock in the ridges. As a result, rivers of the region drain from west to east and suddenly make almost 90-degree turns to the south flowing through narrow water gaps. This is true of the rivers Bandon, Lee, Bride, Blackwater, and Suir. Where the rivers enter the sea, they form excellent harbors.

The central lowlands region begins along the Atlantic Coast between the mouth of the River Shannon on the south and Galway Bay on the north. It extends eastward across the island to occupy area between Dublin and Dundalk. In the north, the central lowlands begin at Laugh Foyle and extend southward to Tipperary and Tralee. The lowlands range in elevation from sea level to 400 feet (122 meters). While the central lowlands indeed reach sea level in several places along Ireland's shores, most points of the lowlands region range from 200 feet (61 meters) to 400 feet (122 meters).

Understanding glacial erosion processes is crucial to under-standing the physical geography of Ireland. With glaciation, the land was bulldozed as the ice moved forward. Later, the glaciers dumped debris as they melted and retreated, resulting in uneven terrain. The terrain of the central lowlands is generally covered by layers of drift materials (rocks, gravels, and clays) in varying thickness.

The northern part of the central lowlands is covered by thousands of glacial hills, called drumlins, and lowlands or hollows. Drumlins are easy to recognize since they resemble the upside down (inverted) bowl of a teaspoon with a steep rounded side and a more gentle tapering side. Hollows are the lowlands between the hills. Drumlins extend offshore along the northwest coast, where they form islands that can be easily recognized by their inverted spoon shape.

The basin of the River Shannon occupies the middle of the central lowlands. This is a land of limestone bedrock covered with glacial drift. The drift is thickest east of the River Shannon. It becomes quite thin on the west where streams have cut into the bedrock, resulting in valleys, lakes, and bogs. In the extreme west, the bulldozing action of glaciers resulted in an area of extensive exposed limestone bedrock pavement called the Burren. Once exposed, the limestone bedrock of the Burren experienced weathering from ice, wind, and rain. This erosion of limestone has created crevices that lead to limestone caves carved below the surface.

When a chemical process below the surface weathers lime-stone, geographers call the landscape karst topography. Karst topography is marked by sinkholes, deep cracks in the limestone, underground drainage, and caves. County Clare has an extensive area of karst topography. Unlike other karst areas around the globe, the ancient Irish did not use caves in the Burren as dwellings.

South of the Burren are the Cliffs of Moher. The Cliffs extend almost five miles (eight kilometers) and stand over

Ireland

- ···-·-·- International Boundary
- ----- County Boundary
- ★ National Capital
- • Major Cities
- ——— Rivers
- ——— Roads

The island of Ireland extends 302 miles (486 kilometers) from north to south and is 171 miles (275 kilometers) wide. The island is about the size of Maine. The Republic of Ireland is similar in land area to West Virginia, and Northern Ireland is equivalent to Connecticut in size. The interior, the central lowlands, is surrounded by highlands; Ireland's highest mountain, Carrauntuohil, reaches 3,414 feet (1,041 meters). The island is surrounded by the Atlantic Ocean and Irish Sea which create a coastline exceeding 3,500 miles (5,633 kilometers).

700 feet (214 meters) above the Atlantic Ocean. They form a massive, yet beautiful, barrier to erosional attacks by the wind and waves of the sea. These cliffs are composed of layers of three sedimentary rocks: sandstone, shale, and siltstone. Of the three layers, sandstone is the most resistant to erosion and forms the protective cap rock, slowing erosion over much of the cliffs.

Although geographically part of Northern Ireland, the island also contains a fascinating landscape known as the Antrim Basalt Plateau. This landscape is a result of repeated horizontal lava flows. The Antrim Basalt Plateau rises over 1,000 feet (305 meters) high. Over time, the plateau has eroded into a series of sheltered valleys. Lough Neagh, the largest lake in Northern Ireland, is located in the plateau. The most spectacular topography in the Antrim Basalt Plateau is the Giant's Causeway. It is an ancient zone of cliffs formed where the horizontal layer of black basalt cooled, shrunk, and resulted in the formation of almost 40,000 primarily hexagonally shaped columns that fit together as if laid in place like a mosaic.

CLIMATE

Weather is the atmospheric conditions at a particular moment in time. Climate is weather averaged over a long period of time. Climatologists explain that one of the greatest influences on climate is the latitude of the place. Ireland is located between 51° and 55° north latitude. It is much closer to the North Pole (90° north) than it is to the equator (0°). Climatologists describe the lands between 35° and 55° as being "middle latitude." Just knowing where Ireland is located, in the middle latitudes, and that it is an island, tells a lot about its climate. The middle latitude location, for example, indicates that the climate will be moderate, not the extreme cold of the poles, or the extreme heat of the tropics.

Islands typically have moderate temperatures because the surrounding ocean does not experience severe temperature changes. The consistent temperature of the Atlantic Ocean helps moderate Ireland's climate. So do the warm waters of the Gulf Stream and North Atlantic Drift (ocean currents). Despite being located at latitudes comparable to central Canada, Ireland's average monthly temperatures remain above freezing during the winter months.

The climate of Ireland is temperate and moist, with an average yearly temperature of 50° F (10° C). Summers are mild with average temperatures of about 60° F (16° C) during July and August. Winters are tempered and modified by the sea, resulting in an average January temperature of 41° F (5° C). Mean daily minimum temperatures average 36° F (2.5° C). Frost and snowfall are still possible, but they are generally short-lived. Frost is relatively rare in western Ireland.

While Ireland's climate is not one of extremes, it is one of rapid change. The high-pressure system over the North Atlantic Ocean consistently brings moist air onto Ireland throughout the year. In the northern hemisphere, winds circulate clockwise around the high-pressure system. Thus, westerly winds are consistently blowing moist air onto Ireland's west coast. Parts of Ireland can experience fog up to 60 days a year. This combination produces a damp, humid climate with consistent precipitation all year long.

Rainfall amounts are greatest along the entire west coast, especially in the mountains. Although monthly precipitation is fairly consistent, generally the heaviest rainfall is in the winter months and lightest is in the early summer. Average annual rainfall is 43 inches (109 centimeters). Actual rainfall, of course, varies from place to place. Dublin has one of the lowest totals, averaging 31 inches (79 centimeters) a year. The heaviest precipitation amounts are generally found in the mountains of the west where the yearly total can exceed 118 inches (300 centimeters) a year.

SOILS

The fertility of soils varies, in large part because the material from which soils are formed (typically the underlying bedrock) also varies across places. This "parent" material is not the only factor in determining soil fertility. Ireland's climate makes the production of soils difficult. With few climatic extremes, it takes longer for vegetative materials to break down and for soils to develop.

Ireland's soil compositions are complicated by the presence of glaciers during the Ice Age. The glaciers pushed and deposited soils and rocks from points north and east onto Ireland. This glacial drift covers the island at varying depths. When a glacier sits in one place and melts over time, the sediment it deposits stratifies into layers of sand, silt, clay, and gravel. Even massive boulders are deposited.

Human activities also alter soil development. The Irish removed the forests, farmed, grazed, and fertilized the land, modified drainage, and dug turf. Hundreds of years ago, the Celts were known for adding seaweed to the soil to try to increase its fertility, just as some gardeners add compost to their soil today.

Of all the ways the Irish have changed the soils, the most interesting and best known is the creation of peat bogs. There are two kinds of peat bogs: blanket bogs and raised bogs. Both kinds of bogs are composed of vegetative matter that is poorly decomposed and has accumulated over time in waterlogged soils.

Blanket peat bogs are located mainly in western Ireland, where they cover many of the rolling green hills. Blanket bogs formed, starting 6,000 years ago, because of deforestation caused by humans. Once the early farmers cleared the land of trees, soil nutrients washed away and the underlying material hardened. This made it difficult for water to be absorbed beneath the layer of hard material. As vegetation died on the hills, the remaining material would simply sit, along with water,

Peat, shown here being cut for fuel, is formed when vegetable matter decays and becomes waterlogged, forming peat bogs. Beginning in the 1600s, the Irish began to cut the peat, dry it, and use it as a fuel. Peat is important to the Irish economy since it is used for heating, as fertilizer, and as a fuel at power generation plants.

in the waterlogged hills. This material became peat. Thousands of years later, in the 1600s, the Irish began to cut the peat, dry it, and burn it as fuel.

Extensive dome-shaped, raised bogs are found in the central lowlands. These bogs can be up to 30 feet deep and cover several square miles in area. Raised bogs were glacial lakes thousands of years ago. Over time, poorly decomposed vegetative material began to fill the lakes. The waterlogged soils lack oxygen, and without oxygen, vegetation cannot decompose easily. Because the vegetative material cannot decompose easily, the bogs have grown higher than their surroundings over time. The most extensive raised bogs are the Bogs of Allen, extending across three counties. Today, peat in the raised bogs is commercially cut and sold for fuel.

FLORA

Over the eons of time, Ireland has been a land of changing climates. With climatic changes came changes in vegetation. Sometimes these climatic and vegetation changes took millions of years. On other occasions, change occurred more rapidly, as happened following the last Ice Age.

At that time, Ireland's flora (plant life) was composed primarily of tundra plants. As the climate warmed, the vegetation changed. Extensive oak forests grew to cover most of the island. In certain areas of higher elevation, the oak forests became mixed with pine and birch trees. In lower elevations, oak mixed with elm and alder. Across the island, under the canopy of trees, plants such as nuts, berries, ferns, and mosses thrived.

In discussing the environment, geographers often emphasize the native plant life. This is really not possible to do in the case of Ireland. Instead of becoming stewards of the land, the Irish completely altered the plant environment. About 6,000 years ago, they began cutting down the oak forests. This was the beginning of a destructive deforestation process that continued for several thousand years.

By 1500 A.D., good forests still remained in Ireland, but during the next two hundred years, virtually all forests were cut to provide lumber for the British to use in shipbuilding, charcoal, and barrel making. By the middle of the 1700s, the oak forests were gone, and Ireland was forced to import wood.

Today, through human action, forests cover about three percent of Ireland. The government, through the Forestry Commission, has established fast-growing coniferous plantations of spruce, pine, and fir. Other trees found in Ireland's remaining forestlands include oak, alder, ash, beech, birch, chestnut, elm, hazel, sycamore, and willow. Because it rarely freezes, even some hardy species of palm trees can be found along the south and west coasts of the island! Some of the large estates of the former British landlords contain examples of individual trees from around the world. Even gigantic, young American Redwood trees can be found in Ireland.

Flora of the mountains and highlands tends to be either blanket bogs or the moorlands of grasses, heather, and bracken ferns. Expanding thickets of gorse are also attacking these areas. Gorse is a dense, spiny, evergreen shrub with yellow flowers that crowds out other vegetation while dying at its own center. Gorse is a parasite to Ireland's plant life today, much as the kudzu vine is in the southern United States.

Ireland is also a land of beautiful flowers. Among the most fascinating are the beautiful pink flower, thrift, the large white sea champion found along the seashore, the yellow fleabane that resembles dandelions, and water lobelia, a plant whose leaves grow below lake waters and whose lilac-like flowers grow above.

FAUNA

In the geologic past, Ireland was once the home of woolly mammoth, tundra reindeer, spotted cave hyena, arctic fox, lemming, and migratory birds. Over time, brown bear, giant Irish deer, wolf, and wild boars replaced the mammals. The deer fell victim to carnivores and the bear and boars disappeared after the arrival of people.

Today, protected wild red deer are the largest animals on the island, but they are not the same species as the giant Irish deer of the past. Today, Irish wildlife includes primarily the gray seal, common seal, fox, badger, weasel, pine marten, Irish hare, rabbits, red squirrel, hedgehogs, shrew, and bats. The island also is home to turtles, frogs, toads, and one species of lizard. There are no snakes! Additionally, several hundred species of birds, both native and migratory, inhabit Ireland. They include the wren, raven, jay, callows, chough (in the cliffs of the extreme west), grebe, gannet, puffin, and whooper swan.

Fish species are numerous and have changed considerably over time. The River Shannon salmon are a popular food on the island. On aquafarms, the Irish raise salmon, turbot, prawns, and mussels. The sea around Ireland is home to pollock, mackerel, herring, sprat, blue whiting, silver smelt, gunner, cod, lobster, and crab.

WATER

The Atlantic Ocean and the Irish Sea surround Ireland, creating over 3,500 miles (5,633 kilometers) of coastline. The wave action from the ocean and the sea work to create beaches and rugged coastlines.

The River Shannon, the longest river on the island, dominates the central lowlands, flowing from County Cavan south through the center of the island. It is 211 miles (340 kilometers) long. The River Shannon widens significantly in places, forming several broad lakes, most notably Lough Derg and Lough Ree. At the city of Limerick, the river turns almost directly west, the gradient steepens, and the river begins downcutting with greater force as it drains through the Shannon Estuary. Today, the River Shannon is one of Ireland's major tourism and recreation attractions.

The second longest river on the island is the River Barrow. It flows south from the midlands toward Waterford Harbour where it is first joined by the River Nore and then the River Suir.

Major lakes on the island include the Lakes of Erne, Lough Corrib, and Lough Ennell. Lough Neagh in Northern Ireland is the largest lake on the island, with a surface of over 153 square miles (396 square kilometers). Most lakes occupy basins scoured by glaciers and filled with glacial meltwater.

In addition to the rivers and lakes of Ireland, in the 1800s parliament funded the building of inland waterways, linking Dublin with the River Shannon. The Grand Canal starts in Dublin and extends westward 81 miles (131 kilometers) to Shannon Harbour. It also contains 30 miles (49 kilometers) of branch canal. A second canal, the Royal Canal, was developed by Dublin businessmen to compete with the Grand Canal. The

Matthew Bridge crosses River Shannon in County Limerick which is located in Ireland's west central lowlands. The river, Ireland's longest at 211 miles (240 kilometers) is important for tourism and recreation, especially salmon fishing.

Royal Canal begins at the River Liffey and extends 90 miles (145 kilometers) west to the River Shannon. The Irish government restored the Grand Canal, and it is upgrading the Royal Canal. These waterways provide Irish citizens and tourists with a variety of recreational activities including boating, fishing, and sightseeing.

MINERALS

Ireland is not a mineral-rich island. It has small amounts of several important minerals. Minerals are generally identified as one of three types: metallic, nonmetallic, and mineral fuels. Metallic minerals have a long history of importance to people. The most commonly occurring metallic minerals in Ireland are lead and zinc. The most significant deposit of these two minerals is at Navan, where mining began in 1977. Ireland is now among the European Union's leading exporters of lead and zinc.

The most abundant nonmetallic mineral in Ireland is limestone (two-thirds of the island is composed of limestone bedrock). Sand and gravel are also important nonmetallic minerals. Numerous stone quarries can be found on the island. Some date back hundreds of years and are no longer operational. Others are recent and fully operational. These materials are used primarily in construction and road building.

One important aspect of the nonmetallic mineral industries is Irish marble. Ireland's limestone bedrock provides the marble industry with access to about 40 different marble patterns and colors. The most famous of the marbles is the green Connemara marble from 600-million-year-old Precambrian deposits.

Mineral fuels are also sparsely located in Ireland. Coal deposits are rare and of minor importance. There are natural gas and oil deposits offshore, and natural gas production off the southwest coast is becoming increasingly important. The most important fuel continues to be peat from the bogs, which is used for heating fuel, fertilizer, and electrical power generation.

An early perception of Ireland is a land of forty shades of green, a country of rolling hills covered with lush green grasses, fenced off with stone and bush fences. Studying the physical geography of Ireland, shows why this perception is accurate.

Ireland's hills, mountains, and cliffs form a rim around the central lowlands of the island. In the west and the southeast, the hills are part of ancient mountain ranges that stretch into Europe. Throughout the hills in the coastal zones are the telltale signs of glaciation, from the drift-covered hills, to the drumlin islands, to glacially deposited fieldstone fencing. In the west, these hills are blanketed with peat bogs, creating multiple shades of green.

In the central lowlands, the hills are typically raised bogs, formed over thousands of years. The River Shannon dominates the landscape and history of the central lowlands.

Ireland's location on the eastern edge of the Atlantic Ocean accounts for its moderate, moist climate. The westerly winds of the middle latitudes consistently carry moist air from the Atlantic over the island, providing much fog and rain throughout the year and ensuring that the island remains forty shades of green—the "Emerald Island."

After the Stone Age people arrived in Ireland thousands of years ago, they built impressive burial mounds such as the Passage Tomb at Newgrange. Located in County Meath which borders the Irish Sea, the Passage Tomb was built around 3000 B.C.

Culture History and Cultural Landscape

adó, fadó, with the end of the Ice Age, 12,000 years ago, the glaciers retreated. They left Ireland covered with glacial debris and tundra vegetation. It was a barren land, with limited varieties of plant and animal life. This is the setting in which Ireland's changing culture history and cultural landscape takes place.

Over the next 2,000 years, Ireland's climate warmed from tundra (sub-arctic) to temperate. As the climate warmed, oak and elm trees grew. Based on existing archaeological evidence, the first human occupants of the island arrived in the north, sometime between 8000 to 7500 B.C. They most likely arrived in present-day Northern Ireland from Scotland.

EARLY PEOPLES

With the arrival of Neolithic (Stone Age) people about 4,000 to 6,000 years ago, came the initial development of a uniquely Irish cultural landscape. Since that time, the cultural impacts of successive peoples, cultures, and generations have all left their imprint on the Irish landscape.

A cultural landscape is the visible imprint of human activity on the land. For example, one could look at the monuments, museums, and government buildings on the Mall in Washington, D.C. These structures indicate that American culture values its history, memorializes important historical figures, and appreciates an ordered layout of buildings and planned spaces. Cultural landscapes are layered, and each successive human group occupying a place changes the landscape to reflect its culture. Changes in cultural landscapes can be seen when people use the land, create architectural styles, construct buildings, and develop their roads and towns differently than the generations who preceded them.

The cultural landscape of Ireland is particularly rich because, as an isolated island with a temperate climate and blessed with stone, it has escaped many of the more destructive attacks of erosion, warfare, and time. In recent years, the Irish government and the European Union have invested millions of euros to restore and preserve the many layers of Ireland's cultural landscape.

When Neolithic people arrived, they began to change the natural landscape by clearing forests to plant grain crops. Clearing forests spurred the formation of blanket peat bogs in western Ireland. Impressive burial mounds, such as the Passage Tomb at Newgrange, are another notable example of the presence of Neolithic people. These tombs provide clear evidence of their presence in the Valley of the River Boyne. Some Stone Age archaeological evidence points to linkages with peoples as far away as Greece.

About 4,000 years ago, Bronze Age peoples from France and the Mediterranean area began arriving on the island. The Irish commonly refer to these people as the Firbolgs. They used copper found on the island, and mixed it with tin (likely from England) to make metal. The metal tools of the Bronze Age quickly replaced Stone Age artifacts. These new arrivals made jewelry, tools, bronze weapons, and beautiful pottery. Examples of these artifacts are in museums across the land.

Bronze Age peoples lived in hill forts, which were small communities surrounded by stockade fences. Over time, the forts decayed, although some outlines of their shape can still be found. They hunted, farmed, and raised the same varieties of crops and domesticated animals found today on Irish farms. Bronze Age wedge tombs and burial pits remain as fascinating parts of Ireland's cultural landscape.

Almost 3,000 years ago, the Iron Age reached Ireland and lasted until approximately 500 A.D. The Iron Age had a phenomenal impact on Ireland and its people. This is the era in which Celtic influences first appeared, about 2000 B.C. Exactly how these influences arrived is unknown, but Celtic designs began appearing in jewelry and weaponry.

THE CELTS

The Celtic people lived as far north as Ireland and as far south as Italy. During the Celtic Iron Age, Ireland's cultural landscape consisted of as many as 200 small kingdoms, each with a ruler. Both men and women served as rulers and warriors in Celtic society. The impact of these small Celtic kingdoms can still be seen in the toponyms (place names) on the map of Ireland today. There were actually three levels of kings and kingdoms. The simplest type of king ruled a small kingdom. The second type of king ruled over the kings of several small kingdoms. The third type of king ruled over provinces that were comprised of the small and combined kingdoms. Over time, four of these provinces ruled by the most powerful kings survived. Today, the names and extent of those kingdoms still survive on maps as the four provinces of Ulster, Connaught, Leinster, and Munster.

The sites and cultural landscapes of the more famous and extensive kingdoms are preserved today as Office of Public Works Heritage Sites in Ireland, or National Trust Sites in Northern Ireland. In County Meath, the Hill of Tara is the most important of these ancient sites. It was the most important royal site during the entire Iron Age and into the Age of Christianity because it was still the seat of the high kings of Ireland as recently as 1,000 years ago. Other major royal sites and forts of the Iron Age include Dun Ailinne (Kildare), Rath Cruachain (Roscommon), and Dun Aonghasa (Aran Islands). The cultural landscape of Iron Age Ireland was covered with *raths* (ring forts). The kings positioned the ring forts on prominent hilltops. Often, the rath enclosed farmsteads, especially cattle and sheep. These forts discouraged cattle raids and theft.

Little is known of the religion of the ancient Iron Age Celts. Druids were important people in the kingdoms. They were somewhat like a combination of priests, judges, doctors, and prophets. It is known that many of the places held sacred to the Celts have maintained their status today, often being adopted as sacred by Christians.

The Celtic Age in Ireland coincided with the age of the Roman Empire in the Mediterranean. The Roman Empire reached as far north as Scotland. The Irish are proud that the Romans never claimed their land. However, Roman culture diffused into Ireland despite the absence of the Roman army.

The most important invasion of Ireland in the last 2,000 years was not military but religious. This Roman-based invasion had more impact on the cultural landscape than any other event in the history of Ireland. It began in 430 A.D., when Pope Celestine, the leader of the Roman Catholic Church, sent

Palladius to Ireland to establish Christianity. Palladius died in 431 A.D. The following year, the most celebrated person in the history of Ireland arrived.

SAINT PATRICK AND CHRISTIANITY

In 432 A.D., Saint Patrick arrived in Ireland. Saint Patrick was born in Scotland in 387 A.D. and died at Downpatrick, Ireland, on March 17, 493 A.D. According to these dates, Saint Patrick would have lived to be 106 years old.

The common story is that slave raiders kidnapped Saint Patrick when he was a young boy and sold him to an Irish chieftain. He then tended sheep for six years. He learned the Celtic language and culture, but eventually escaped slavery and returned to Great Britain. There, Saint Patrick joined the priesthood and returned to Ireland as a missionary in 432 A.D.

The Irish still tell stories of his experiences and miracles. It was during Celtic times that poets in Ireland developed the oral tradition of verse and storytelling. This tradition continues in Ireland today. In Ireland, it is sometimes difficult to distinguish fact from fancy. Through the storytelling tradition, fancy can become as real as fact. The stories of Saint Patrick are an excellent example of this.

In the United States, there is the story of Saint Patrick driving all of the snakes out of Ireland. In Ireland, the story most commonly heard about Saint Patrick is his confrontation with the Celtic high king on the hills of Slane and Tara in the Valley of the River Boyne. Fadó, fadó, the high king of Ireland had summoned all the kings and chieftains to Tara for the celebration of a great feast. Saint Patrick saw that as an opportunity to meet Celtic leaders, confront the powers of the Druids, and spread the gospel of Jesus Christ. On Easter Sunday, Saint Patrick and his followers proceeded to the Hill of Tara. The Irish say a series of mystical confrontations took place between the saint and the Druids. Patrick prevailed, and the high king gave him permission to preach Christianity in Ireland.

Saint Patrick returned to Ireland in 432 A.D. as a Catholic priest after, according to legend, having been held a slave by an Irish chief. While a slave, he learned the Celtic language and culture which helped in his mission of converting the Irish from paganism. Saint Patrick is the most influential person in Irish history. He brought Christianity to Ireland, established churches and monasteries, and built the framework for monastic societies that became the first Irish centers for learning and literature.

No person, artifact, feature, or event has had more impact on the cultural landscape of Ireland than did Saint Patrick. He established the Roman Catholic faith in Ireland, constructed churches and monasteries, and set the foundation upon which monastic societies were built. Evidence of Saint Patrick's impact on the cultural landscape is virtually in every corner of the land, in churches, sacred wells, schools, hospitals, organizations, feasts, and celebrations.

For the next three or four hundred years, Christianity developed in Ireland. During that time, a new Irish culture also developed because outside influences were few. In 455 A.D., Saint Patrick built the church at Armagh. Today, Armagh remains the island's predominant religious center. It is home to major religious leaders and the cathedrals for the Roman Catholic Church and the Church of Ireland.

Saint Patrick's followers, such as Saint Ciaran, Saint Kevin, and the famous Saint Brendan the Navigator (who some historical geographers believe may have reached North America long before Christopher Columbus) founded monasteries such as Clonmacnoise, Glendalough and Armagh. These monasteries became the Irish centers of learning, literature, and knowledge. In fact, as the dawn of the Dark Ages arrived, these monasteries were the only centers of learning freely operating in Europe. Irish monasteries flourished for over a thousand years despite various attacks. Their missionaries spread Christianity to many lands. They influenced religious art and artifacts. They produced the beautifully hand-illustrated books of Kells, Armagh, and Darrow, which contain the four gospels of Christianity. The Irish monasteries became a major religious, political, cultural, and economic force in Europe.

THE VIKINGS

With the approach of the ninth century, Ireland began experiencing an influx of Scandinavian people, generally referred to as the Vikings, or the Danes as the Irish called them.

Their arrival in Ireland was not unique because they were also raiding and occupying lands as far east as Russia. In Ireland, the Vikings found food and wealth. Food was found across the countryside, and wealth was concentrated in the monasteries. The Vikings were pagans for whom objects used in the practice of religion had only monetary value. The real threat of Viking attacks is still very evident. Many monasteries built tall, round towers to protect their wealth from the Vikings. Over 100 of these towers can still be seen on Ireland's cultural landscape.

The Vikings also modified the cultural landscape. In 841, they built a fort where the River Liffey and River Poddle join. The junction of those two rivers was called a "blackpool" (Dubh Linn), hence the name Dublin. That site is occupied today by Dublin Castle. On the west coast, the Vikings also established the city of Galway where they built a protective wall around the settlement to keep the Irish out. The Vikings brought urban life to Ireland. Before their arrival, the Irish did not live in cities and towns. Vikings also developed forts in harbors where forests provided wood and rivers provided access to the interior.

The Vikings established trade and commerce in the forts, and outside the walls they began farming. They also reintroduced the Irish to weapons and warfare. Soon after the Vikings arrived, the Irish were making swords and building ships.

In the early tenth century, more Vikings began arriving in Ireland. They also raided the land and this time tried to occupy it. In the northern part of the island, the Irish were able to resist those attacks. Along the east coast and in the south, the Vikings began controlling the land. One major exception was in a portion of the Province of Muenster ruled by Brian Boru. He was a warrior king who wanted to rule all of Ireland. He successfully led his forces against the Vikings. In 1014, they drove the Vikings from Ireland, although Brian Boru died in battle. His sons then fought over who was to be high king. In the end, his grandson Turlough O'Brien became high king of all of Ireland in 1072.

BRITISH RULE

In 1169 A.D., Henry II, King of England, sent Anglo-Norman armies under Richard Strongbow to help Dermot MacMurrough, King of Leinster, regain control of his lands. Instead, Henry II claimed all of Ireland for himself. This began the British colonization of Ireland and over 800 years of conflict that resulted from that occupation.

Once in control, King Henry II gave Anglo-Norman nobles control over about two-thirds of Ireland. They again modified the cultural landscape. They built stone castles, maintained powerful armies, and consolidated ownership of vast land areas. Among these nobles were Catholic people of English, French, Flemish, and Welsh origin. The family names of these nobles included FitzGerald/Desmond, Costello, Gautier, deBurgo/Bourke, deCourcy, deLacy, Burton, and Butler. In Ireland today, some people still refer to their descendents as "the Normans."

The British retreated to Dublin to focus on economic and political control of Ireland. This move soon changed the economy and the cultural landscape. The area around Dublin, known as the Pale, was the focal point of British colonialists. Here, the British developed castles and manor houses on large estates that often included designated living spaces for their Irish employees. They brought in people from Britain to control the estates and towns, thereby marginalizing the Irish. They established market towns along the major transit routes to get their products to European markets.

Religious centers shifted from monasteries to more Anglo-Norman-based medieval abbeys and parishes such as St. Mary's, Holy Cross Abbey, and the Rock of Cashel. The abbeys controlled vast acreage and even villages. The Anglo-Normans also used the Irish to construct stone fences and walls around the lands. Actually, stone fences and walls have been part of the cultural landscape of Ireland since the Neolithic Period. Today, there are over 250,000 miles of stone fence and walls in Ireland.

In 1533, King Henry VIII of England broke from the Roman Catholic Church. He established the Church of England and dissolved the Catholic monasteries. In Ireland, Henry VIII's church is known as the Church of Ireland. Henry VIII's actions took place within the context of the Protestant reformation in Europe. Henry VIII was the first monarch to form his own church and declare himself ruler of his subjects' spiritual and temporal (secular) lives.

In the 1600s, the British began to build plantations in Ulster province. They brought loyal Protestants from Scotland and England to occupy the farm lands. Irish nationalists resisted and General Oliver Cromwell, a merciless British soldier, landed armies in Ireland to punish them. They killed thousands of Irish people and burned churches, abbeys, and monasteries in an effort to destroy Catholicism. The ruins of these religious centers are found across Ireland.

By 1688, much of Europe was at war and Ireland became one of the battlegrounds. The armies of Catholic King James II of England and William of Orange, a Protestant from Holland, met at the Battle of the Boyne in June 1690. William and his armies defeated James II and his troops. The treaty following this battle enacted discriminatory laws against Catholics and Presbyterians.

THE GREAT FAMINE

The nineteenth century was an especially difficult period in Ireland's history. From 1845 to 1848, potato blight struck the most significant food crop on the island and the Great Famine ensued. Over 8 million Irish were dependent on potatoes as their primary food source. By 1846, the people were starving and many died from hunger. In weakened conditions, people also died from other diseases such as

In an illustration from a London newspaper of the day a starving boy and girl rake the ground for potatoes to eat. The Great Famine descended upon Ireland in the mid-1840s when the most important food crop, the potato, was struck by blight. In 1846, with hunger, starvation, and death rampant, Irish people started migrating to North America, Australia, and New Zealand to find a better life. In five years, Ireland's population fell from 8 million to 6 million as the result of death or emigration.

typhus and dysentery. The solution to one's survival was to emigrate to America, Canada, or Australia. In five years, the population declined by death or emigration to fewer than 6 million people. In parts of Ireland, the cultural landscape still carries the scars of this disaster, reflected in row after row of abandoned potato mounds.

IRISH INDEPENDENCE

From 1876 until recently, Ireland faced hard economic times. It has only been within the last 25 years that the standard of living for the average Irish person began to improve. During the last 125 years Ireland experienced a change that led from servitude to independence. This change came about as a result of a combination of rebellion, negotiation, conflict, civil war, death, and peace. The cultural landscapes of this process vary. The sites of the rebellion at the Post Office building in Dublin, to Bloody Sunday at Croke Park, the independent Dáil Erieann (Irish Parliament), intellectual centers at University Colleges in Dublin, Galway, and Cork. The literary and theatrical geniuses symbolized by Coole Park near Gort mark this era.

In reality, nowhere in Ireland better symbolizes the culture history and cultural landscape of Ireland in the last 125 years than does the Glasniven Cemetery in Dublin. Here rest the remains of 1,500,000 people. Included are heroes of independence and revolution, politicians, religious leaders, writers, saints, as well as commoners.

The monument to Daniel O'Connell, the leader of Catholic emancipation, dominates the entrance to Glasniven. Nearby are the graves of several heroes of Irish independence. Charles Stewart Parnell, leader of Irish Nationalists who in 1876 fought the eviction of Irish farmers and called for an independent Irish Parliament, is buried in Glasniven. Michael Collins, a true revolutionary, a powerful figure in various Irish republican organizations, and the 1921 signer

4

Irish People and Culture

adó, fadó, Ireland was populated by people from the Eurasian landmass. Racially, the Irish are linked to Western Europe and Scandinavia. The Celtic culture came to envelop Ireland during the years of Celtic domination. Saint Patrick and Christian missionaries brought the Catholic religion to Ireland. Vikings intermarried with the Irish resulting in some genetic changes. Later, the Anglo-Normans brought a new economic structure to Ireland, changing the island's cultural composition in many ways, but changing the racial composition very little. During British colonization, there were more changes in culture, especially a move away from the Irish language to the English language. During the rise of independence, the Irish reached back to their Celtic roots and reaffirmed their Catholic faith in forming the Republic of Ireland.

IRELAND'S POPULATION

One of the most common adjectives people use when describing Ireland is "rural." Ireland, historically, has been a land of scattered farmsteads with few cities and villages. Today, the Irish Central Statistics Office reports that the greater Dublin area has a population over 1 million. Cork is the next most populous urban area, with over 400,000 people. Between 1996 and 2002, Ireland's population grew 8 percent. Each of Ireland's four provinces grew, with Leinster province (home to Dublin and several other cities) growing the most at 9.4 percent. Several cities in Ireland experienced population increases greater than 10 percent in the last six years. The sharp gain is mainly due to Ireland's economic boom. According to Ireland's Central Statistics Office, of the eleven cities experiencing population growth of more than 10 percent, nine are located in the Leinster province. The other two are County Cork in Munster province and Galway city in Connacht province.

Geographers who study population recognize that poor, agricultural areas typically have high birth rates and high death rates. This is because the population desires large families to help work the land, but has little access to health care that could increase life expectancy. When an economy becomes more prosperous (typically more urban as well), the population grows. This is due in part to the decline in death rates resulting from better access to health care.

As the economy continues to gain wealth, the population growth slows. This is because people desire fewer children as they change jobs from agricultural to industrial and service economies. Finally, some countries reach a state of zero population growth. The death rate becomes higher than the birth rate. At the point of zero population growth, the population of a region starts to "gray," meaning a larger proportion of the population is over retirement age. This places a burden on the smaller population of working-age people who must support the retirement of the larger, older generation. Today, Italy and Sweden are two countries with this burden.

This explanation of population change depending on birth and death rates does not account for the growing population in Ireland today. Ireland's recent population growth does not stem from a higher birth rate than usual. Rather, Ireland's birth rate has declined to 1.87 births per woman in the 1990s, from 3.96 births per woman in the 1960s. This is well below the natural replacement rate of 2.1 births per woman. The Irish Central Statistics Office attributes the recent growth in Ireland's population to in-migration rather than an increased birthrate. Between 1996 and 2002, every single province, county, and city in Ireland received more in-migrants than out-migrants. With Ireland's high technology economy creating thousands of new jobs each year, well-educated Irish who formerly left the country are returning home to find work.

The growing computer and pharmaceutical sectors of the Irish economy are locating their companies throughout the country. This accounts for the large in-migration in every province, county, and city between 1996 and 2002. Certainly, many foreign companies chose to place their plants in the greater Dublin area. Others, such as Apple Computer, have chosen Cork, the country's second-largest city. Some foreign companies deliberately choose rural locations. They hope to appeal to the young Irish who grew up in rural counties, are educated in computers and trade, and wish to stay in their rural home areas. For example, Prudential Company set up a financial services operation in rural County Donegal in the northwest of Ireland, hoping the Irish who graduate from the county's schools will stay in Donegal.

Ireland today is a country of one major city, several small cities and towns, a multitude of villages, and numerous farms. The settlement landscape varies throughout the country. In the west, one sees hundreds of farmsteads on oddly shaped fields, reflecting the strength of farming in the region prior to the potato famine. Ruins of cottages and stone fences in the west reveal that the region was much more densely populated prior to the famine than it is now. In fact, today's population of 3,917,336 is still only about one-half of the island's 1845 population.

THE TRAVELERS OF IRELAND

If one used race as the only indicator of diversity, Ireland would be a homogeneous island. More than 99 percent of the island's population is white. Only an estimated 20,000 people are of non-European origin. An additional 22,000 "Travelers" live in Ireland. Travelers are a people who are racially the same as the majority Irish, but are ethnically distinct.

Even though Ireland is racially homogeneous, it is ethnically diverse. Ethnicity is a difficult term to define because it envelops so much. The best way to understand ethnicity is to ask people how they would identify themselves. If all the people see themselves as part of the same group, they are of the same ethnicity. If a group of people focus on some aspect of their culture (economic or political practices, language, religion, role of family, ancestors, economic class, social hierarchy, or traditions) as being distinct from the rest of the people around them, that group of people is an ethnic group.

The Travelers of Ireland are an ethnically distinct group of Irish people. They identify themselves with the Gammon language, with genealogical linkages that show they have lived together for centuries, with a tradition of traveling to make a living, and with many traditions that set them apart from the rest of the Irish. Historically, the Traveler life has been one of movement. They travel through Ireland to make a living through the roadside sale of goods and through seasonal work. Traditionally, Travelers have camped in rural areas of Ireland.

Since the 1960s, the Irish government has encouraged Travelers to live in houses in urban areas and to use designated camps developed by the government in their travels through Ireland. In other cases, their camps can still be seen along public roadways across the land. The Travelers continue to travel from these encampments and from their homes, especially in the warmer months of the year. Travelers choose to travel for a variety of reasons. Most of all, it is what they know. Their culture and traditions stem from centuries of a traveling lifestyle.

Travelers, and many who study their ways of life, argue that it is healthier to travel than to settle and remain in one place. The Traveler lifestyle does not fit the way of life that most Irish and, in fact, most of the world's other people, know: living and working in one place. As a result, some members of the larger Irish population misunderstand the Travelers and harbor prejudices toward them. Other members of the larger Irish population work to help the Travelers uphold their traditions. Still others are unaware of the Travelers' plight in Ireland.

CATHOLICISM IN IRELAND

One common perception of Ireland is that it is a strongly Catholic country. This perception is accurate, but that faith is facing challenges as Ireland as a whole changes. The Roman Catholic Church is a powerful presence in Ireland. Almost 92 percent of the people identify themselves as Roman Catholic. Less than 3 percent of the population belongs to the Church of Ireland.

During the last 300 years of Irish history, the British colonists worked to diminish the role of the Catholic Church and Catholic identity among the Irish. This colonial policy backfired on the British, and the Irish clung more tightly to their Catholic identities as a way of distinguishing themselves from the British. During the movement for independence in the late 1800s and the early 1900s, the Catholic identity strengthened further. When the Irish gained their independence in the 1920s, they created a constitution that reflected the rights and norms of the Catholic Church in a multitude of ways. Catholicism is not the official religion of the country, but a quick study of the first laws and constitution in Ireland demonstrates that the

Although facing challenges as the country itself changes, the Roman Catholic religion is a major force in Irish daily life. Its strong influence can be seen in Ireland's constitution, the country's many sacred sites, the educational system, and medical facilities. Approximately 92 percent of the Irish people are Roman Catholic.

Catholic Church influenced the laws the Irish enacted.

One can see the influence of the Catholic Church in many aspects of Ireland beyond the laws and politics. For instance, Catholic churches, cemeteries, and sacred sites can be seen throughout Ireland's cultural landscape. Geographers who study religion often start their study of a place and its religion by looking at the sites the religion holds sacred. They may also study the pilgrimages the religion's followers make to access their sacred sites. Sacred sites are places that people either revere or fear because something significant to their religion happened in that place. Someone significant may have been born or died in that place, or the uniqueness of the place may bring reverence or fear to the people.

Ireland is a land of many sacred sites. The location and type of sacred sites in Ireland differ from those on the European mainland. On the continent, people often built sacred sites by putting a relic of a saint in a town and building a church around the relic. There, sacred sites tend to be in the midst of urban areas and relatively accessible. Ireland's sacred sites tend to be located in areas of rough terrain. This is because in the Celtic tradition, sites became sacred for their unique physical geographic features. When the Catholic Church took root in Ireland, it adopted many Celtic sacred sites.

A visit to Ireland's most important sacred sites would include Knock Shrine in County Mayo where Catholics believe the Virgin Mary and others appeared in 1879. Among other sacred sites are Croagh Patrick, Ireland's holy mountain, and Clonmacnoise Monastery.

While some scholars say that the new economy of Ireland has lessened the role of the Catholic Church, it continues to operate most schools and many medical facilities, and to provide many social and cultural services. Historically, the Church has provided for many of Ireland's most destitute through its many charities.

Western Europe has experienced a large decline in church attendance over the last century, as societies have become more secular. Observers point to many changes in Ireland's laws in the last decade, including the legalization of divorce, as evidence that the influence of the Catholic Church is waning. There is no doubt that Europe has become more secular and that the role of the Catholic Church in Europe has declined in the last century. It is too soon, however, to say the same for Ireland. Polls show some decline in church attendance among the Irish, but they also show the Irish people retain strong beliefs in the teachings of the Catholic Church. Ireland's Catholic identity is not official. Rather, it is tied into Ireland's independence movement, its culture of music, dance, and literature, its education system, and its everyday life.

Ireland's distinct Celtic language, called Irish, is another cultural attribute that distinguishes Ireland from the rest of the European region. In the constitution, the country recognizes the Irish language as the national and first official language. The constitution recognizes English as the second official language.

THE IRISH LANGUAGE

Celts brought the Irish language to the island, and the British brought English during the colonial period. During the colonial era, the British tried to suppress the Irish language. Many Irish learned English so they could work in the businesses the British brought to the island, or so they could do seasonal work in the United Kingdom. Since the British focused their colonization of Ireland on the Pale, the area around Dublin, the west of Ireland became the stronghold of the Irish language. The potato famine of the mid-1800s threatened the Irish language. Many people in western Ireland died in the famine and many others from that region migrated to North America. Diffusion (spread) of the English language into the island and the deaths of thousands of Irish speakers combined to threaten the Irish language. The 1891 census reported that 85 percent of the people spoke only English.

Once Ireland gained its independence and set up its government, the country started to support the Irish language through its schools and its hiring policies. Teachers taught the Irish language to students, and the government required its employees to be proficient in Irish. However, people continued to speak English at home and in the workplace.

The government changed its policies and began to focus on the Irish-speaking areas of the island, defining them as the Gaeltacht areas, including Donegal, Mayo, Galway, Kerry, Cork, and Waterford. In the 1980s and 1990s, the Irish government encouraged companies to invest or even locate in the Gaeltacht

Street signs reflect Ireland's first official language, Irish, and the second official language, English. Due to earlier English rule and emigration, Ireland lost many of its native Irish speakers. Although the language is making a comeback, only 5 percent of the Irish people report that Irish is their primary language today.

areas. The government hoped new businesses there would keep people in the counties and keep the Irish language alive. The government successfully encouraged families in the west to house children from other regions during the summer months, so the children could be exposed to the Irish language in a home setting. Irish grammar schools also help to promote the Irish language. All subjects are taught in Irish, and the children's parents are encouraged to learn the Irish language and speak it at home.

For visitors to Ireland today, one of the first exposures to the Irish language would be through television. In the 1990s, the government started funding a television station that broadcasts programs in the Irish language. On average, over two million people a month watch the Irish language station. Many follow the popular Irish language soap operas, learning the language through television. Despite these efforts to restore the Irish language, only 5 percent of the people actually use Irish as their primary language today.

Most Americans perceive Ireland to be a homogeneous culture. This is due, in large part, to the media through which Ireland is portrayed. The stories seen or heard of Ireland in movies, television, and books are commonly rags-to-riches stories of immigrants who had a tough go, but somehow found their way. The Irish are one of the most popular immigrant groups in America, even among those with little or no Irish ancestry. The numerous Saint Patrick's Day parades throughout the United States, from Boston and New York to Sioux Falls, South Dakota, attest to America's fascination with Irish culture.

CULTURAL PRACTICES: MUSIC, DANCE, AND LITERATURE

Three popular aspects of Irish culture that have diffused beyond Ireland and captured the interest of many Americans are music, dance, and literature. These three aspects of Irish culture offer people outside of Ireland other opportunities to learn about the country and to create their own perceptions of its land and people. Irish music carries a strong Celtic influence. In recent decades, however, major music producers have helped transform Irish folk music into popular music with listeners around the world. Traditional instruments include the bodhran (drum), harp, tin whistle, flute, and uillean pipes. These are supplemented today by guitar, bass, and synthesizer. The Chieftains are the most well-known Irish music group outside of Ireland. Other popular acts whose roots are in the Celtic music tradition are Gael Linn, Clannad, and the Corrs. Celtic music has a spirituality to it that attracts many listeners and is especially popular among people of Irish ancestry in North America.

Irish music is popular worldwide. Heavily influenced by Celtic music, traditional Irish folk music has been transformed into popular music by leading musical groups and producers. Among the most popular instruments to be seen at a local pub (bar) are the Irish harp, bodhran, tin whistle, and guitar.

On any Saturday night in Ireland, at many a local pub, one can see Irish dancers perform. Outside of Ireland, the 300-year old tradition of Irish dance is gaining in popularity and can be seen in productions such as *Riverdance* or *Lord of the Dance*, in local Irish festivals, and at Saint Patrick's Day celebrations. Irish dance can be performed solo or in figures (groups). The dances include jigs and reels, hornpipes, and the set dance. During the independence movement in Ireland in the late 1800s and early 1900s, Irish dance became more popular in Ireland. The Irish embraced Irish dance then as another way to distinguish themselves from the British. It helped them take pride in their Irish culture. In the early 1900s, Irish dance diffused and became a popular way for people of Irish ancestry to reconnect with their roots.

The Irish cultural tradition with the widest global influence is Irish literature. Irish poets, playwrights, and novelists have a long history of followers throughout the English-speaking world. Among Ireland's most famous poets are Oscar Wilde, Samuel Beckett, Seamus Heaney, and William Butler Yeats. Among Ireland's most famous writers are Padraic Pearse, James Joyce, Sean O'Casey, Lady Isabella Augusta Gregory, and Brendan Behan. Many of these famous Irish writers lived and wrote in the late 1800s and early 1900s with their popularity and prolific writing tied to Ireland's movement for independence. The stories of their day can be read in many of their poems, plays, and novels.

SUMMARY

Race, ethnicity, religion, language, and cultural practices like music, dance, and literature are common ways of identifying a culture. Geography recognizes that cultures change over time, and in Ireland over the last 30 years, a new identity has started to develop. Some people in Ireland (and the European Union) are starting to see themselves as European. In a recent poll by *Eurobarometer*, 72 percent of the Irish people surveyed said Ireland's membership in the European Union was a good thing. Among the 15 European Union countries, 48 percent of the people also said membership was a good thing. Ireland and Luxembourg tied as the countries offering greatest support for the European Union.

This strong support among the Irish is founded in the benefits the Irish have received from their membership in the European Union. Among the member countries, the Irish had the highest positive response to the question of whether or not their country has benefited from membership in the European Union. Fully 83 percent of the Irish polled believed that Ireland has benefited.

As the Irish continue to benefit from their membership in the European Union, they may increasingly see themselves as

European first and Irish second. In fall 2001, 38 percent of the Irish who responded to the *Eurobarometer* poll said that in the near future they may see themselves as being both Irish and European. Three percent said European first and then Irish, and 2 percent said European only. The largest proportion, 55 percent, saw themselves as Irish only. What all of this means is that the Irish are still trying to figure out what being members of the European Union means to them. They are still wrestling with who they are and where they "fit" in terms of being citizens of Ireland, Europe, and the world.

Studying Ireland's cultural geography reveals the many ways its interactions with the Celts, Christian missionaries, Vikings, Normans, the British, descendants of Irish migrants, and the European Union have affected the traditions and practices of the Irish people. The racial and ethnic composition of Ireland reflects the presence of all these peoples over the last 2,000 years. Ireland's religious traditions reflect the influence of the Celtic religion, the Catholic Church, and the British policies against the Catholic Church. The language reflects the presence of the Celts and the British, as well as ramifications of the potato famine, the great migration, and the policies of an independent government.

The cultural practices of music, dance, and literature reflect the Celtic tradition. They also are traits of a people seeking to become independent and the desire of people outside of Ireland to retain their Irish roots. Finally, the role of the European Union in Ireland promises to lead to changes in the cultural geography of Ireland's future.

The Dáil Éireann (House of Representatives), shown here in session, is one of two legislative groups in the Irish Parliament. The other, the Seanad Éireann, is the Irish equivalent of the United States Senate. The 26 counties of Ireland are divided into 41 constituencies for representation in the Dáil Éireann. All representatives in the Dáil Éireann are elected directly by the Irish people.

5

Government and Politics of Ireland

F adó, fadó, the Celts ruled Ireland as a multitude of kingdoms, with one overall, high king. When the Anglo-Normans came to Ireland, they ruled the area within their own kingdoms, building castles throughout the land. During British colonization, the Irish had little control over their land. As the colonizer, the United Kingdom ruled the land from their capital city in London.

BRITISH COLONIAL RULE

During colonization, Catholics suffered extreme discrimination. The British took lands from the Irish and redistributed them in large landholdings among Protestants who were loyal to London. By 1703, Catholics owned only 14 percent of the land in what is today Ireland. In today's Northern Ireland, Catholics owned only 5 percent of the land. The land ownership pattern in the north resulted from plantation landlords and Protestants brought in from England and Scotland to occupy the area.

The British had a different set of laws for Catholics and for Protestants. Laws that applied to Catholics were called penal laws. Among other restrictions, the penal laws prevented Irish Catholics from voting, from serving in the military, from serving as members of parliament, from carrying weapons, from buying land, and from running Catholic schools. In 1829, the British passed the Roman Catholic Relief Act that repealed the penal laws.

Between Britain's penal laws and their efforts to stop the Catholic Church from operating in Ireland, Irish Catholics suffered greatly during the colonial period. By the late 1800s, many Irish Catholics became anxious for independence and started to make movements for freedom from British rule.

However, not all Irish wanted independence. Most of the Protestants, especially in the large Protestant settlement in the north of Ireland, wanted to remain part of the United Kingdom. The majority of people with close ties to the United Kingdom were descendants of British colonial migrants. Those loyal to the United Kingdom became known as Unionists, and they were mainly Protestants.

HOME RULE AND IRISH INDEPENDENCE

In the late 1800s, a group of Irish Catholics began to demand home rule. Their first goal was to create an Irish Parliament that would have control over domestic issues for the entire island. Members of the Home Rule party became known as the Irish Nationalists. Politically, there were three major interests in Ireland: the Unionists, the Nationalists, and a third group emerging about this time, the Irish Republicans. The Irish Republicans wanted a free Ireland.

They wanted the entire island to gain its independence and exist as one country ruled by Irish.

At the end of the 1800s, William Gladstone, Prime Minister of the United Kingdom, decided to support the Nationalist position and attempted to restore an Irish Parliament on the island. Unionists resisted Gladstone's attempts. The Republicans continued to demand freedom for the entire island and, in 1905, Arthur Griffith formed a new Irish Republican political party that he named Sinn Féin.

In 1912, the Parliament in Britain introduced another bill for Irish Home Rule. Unionists fought the bill even more than the earlier versions. As a compromise, Britain proposed that the Home Rule bill exclude six of the counties in Ireland's northern Ulster province. In 1913, a group of Protestants in the north formed the Ulster Volunteer Force with a goal of maintaining a separate, Protestant Ulster.

Both the Ulster Volunteer Force and the Irish Nationalists began to obtain arms. London's attention turned from Ireland to the rest of Europe in 1914 when World War I began. During the war, both Unionists and the Nationalists fought for Great Britain.

At the same time, Nationalists and Republicans began planning a revolt against the United Kingdom. The uprising occurred on Easter Sunday 1916 when 1,500 rebels occupied key buildings in Dublin. They seized the Post Office, raised the Irish flag, and declared Ireland independent. The British attacked the rebels and killed almost one-third of them. The British then executed 97 of the rebels including Patrick Pearse.

Because of the uprising, more Catholics in Ireland began to support Sinn Féin, Eamonn de Valera's political party that supported Irish independence. In the 1918 election, Sinn Féin won 73 of the island's seats in the British Parliament.

Instead of going to parliament in London, the members of Sinn Féin convened their own parliament in Ireland in 1919. They named their new legislative body the Dáil Éireann

(Irish Parliament). The Dáil Éireann drafted and signed an Irish declaration of independence. The Dáil proclaimed, "We solemnly declare foreign government in Ireland to be an invasion of our national right which we will never tolerate, and we demand the evacuation of our country by the English Garrison."

In 1919, the Irish Republican Army, led by Michael Collins, fought a war for independence. In 1920, the British passed the Government of Ireland Act partitioning the island into north and south. The six counties of the north, a Protestant stronghold, formed their own parliament, and the 26 counties of the south had their own parliament. Under this act, both the north and the south remained part of the United Kingdom.

The province of Ulster in the north includes nine counties, but only six of the nine became Northern Ireland. Protestants in the north recognized that because of the distribution of Catholic and Protestant residence, they could have a stronger majority in their parliament if they only included six of the counties.

Despite clear opposition from the Catholic Church, the Irish Republican Army carried on a guerrilla war. Black (police) and Tan (army) British troops, so named because of the color of their uniforms, fought back mercilessly, also killing civilians.

On Sunday, November 21, 1920, the Irish Republican Army killed 14 British spies. The Black and Tan responded by firing into a crowd of civilians at a football match in Dublin's Croke Park. Twelve people died and this event became known as Bloody Sunday. The Irish Republican Army responded by killing more British soldiers.

Sinn Féin continued to boycott the British Parliament and kept meeting as the Dáil Éireann. The Irish Republican Army continued the guerrilla war and the British continued to respond, but finally, both sides began to talk to each other.

In 1919, Michael Collins led the Irish Republican Army in Ireland's war for independence from Great Britain. Later, in 1921, Collins signed a truce between the Irish Republican Army and the British, creating the Irish Free State. Collins was killed in 1922 during the Irish Civil War.

THE IRISH FREE STATE

In 1921 Michael Collins signed a truce between the Irish Republican Army and the British. The Anglo-Irish Treaty created the Irish Free State. As a free state, Ireland would remain within the British Commonwealth, but would no longer be part of the United Kingdom. The British government became the United Kingdom of Great Britain and Northern Ireland. Some Republicans and Nationalists saw Collins as a traitor to a unified Ireland. History seems to show him as a realist.

In 1922, Eamon de Valera became president of the Irish government. The new government was split between those loyal to Collins and those loyal to de Valera. De Valera and his loyalists walked out of the government. Events led to the Irish Civil War in 1922 and 1923. During that war, Michael Collins was killed in County Cork.

The public demanded an end to the war, but political conflicts continued in the Irish Free State. The political conflicts made it difficult for the government to accomplish anything. In 1927, de Valera and his followers formed Fianna Fáil, a new political party that quickly received a lot of support. By 1932, this new party controlled the Dáil, with de Valera serving as Taoiseach (Prime Minister). The new government enacted several reforms and a new constitution.

THE CONSTITUTION AND THE REPUBLIC

In 1937, Ireland established its constitution. Today, Ireland still struggles with a desire to peacefully reunite Ireland and Northern Ireland into one country. In their constitution, the people of Ireland dealt with the political split of the island by extending Irish citizenship to all people born on the island of Ireland, including the six counties of Northern Ireland. The Irish constitution also expresses a desire to reunite the north and south and to do so peacefully.

In 1949, Taoiseach John Costello declared Ireland to be the Republic of Ireland, a fully independent country. The British finally recognized its status and also gave Northern Ireland the right to decide its future. It remained in the United Kingdom.

THE IRISH GOVERNMENT

In their constitution, the Irish set up a parliamentary government. The Irish government is divided into three separate powers: the legislative, executive power, and judicial.

In Ireland's parliamentary system, the president plays a different role than in the United States. In the United States, the president is part of the executive branch. In Ireland, the president is part of the legislative power, and only a small part. In Ireland, the main duties of the president are to appoint the Taoiseach, to dissolve the Dáil and call for new elections, to command the defense forces, and to sign bills into law. The president serves a seven-year term and may serve two terms.

The focus of the legislative power is the Oireachtas (National Parliament). The Oireachtas is composed of a Dáil Éireann (House of Representatives) and the Seanad Éireann (Senate). The president can dissolve the Dáil and call for new elections. When the president calls for a new election of the Dáil, an election must be held within 25 days. This short time span between the call for an election and the election makes the campaign for Ireland's Dáil quite different from the lengthy political campaigns in the United States.

The government divided the 26 counties of Ireland into 41 constituencies for representation in the Dáil. Each constituency has at least three representatives in the Dáil. Some constituencies have many more because they have higher populations. The Dáil is the body of Oireachtas that represents the people. The people of Ireland directly elect members of the Dáil. All bills in the Oireachtas that involve allocating money must originate in the Dáil.

The Seanad is composed of 60 members. Ireland designed the Seanad to represent the country's different interests. The Taoiseach nominates 11 members of the Seanad. Graduates of Irish universities elect six other members of

Counting votes in an Irish general election is different than the system used in the United States. Ireland uses a proportional representation system in which the Irish voter ranks his candidates on the ballot (first, second, third, etc.) as opposed to the American way of casting a single vote for just one candidate. When the votes are counted in Ireland and a voter's first choice already has enough votes to win, the voter's vote goes to his second choice. The purpose is to make sure minority parties have representation in the government.

the Seanad. The remaining 43 are elected from five panels of candidates. The Seanad's main functions are to consider legislation, make amendments, and pass or reject each bill sent on from the Dáil.

In the executive power, the Taoiseach (Prime Minister) is the head of Ireland's government. Members of the Dáil nominate the Taoiseach, and the president officially appoints the Taoiseach. The Taoiseach represents Ireland in its relations with other countries. Domestically, the Taoiseach coordinates Ireland's 15 departments of state and chairs meetings of the cabinet. As a member of the executive branch, the Taoiseach oversees the execution of the laws the Oireachtas passes.

In the judicial power of the Irish government, there is a supreme court, a high court, and several other courts. All are responsible for interpreting and applying the laws the legislative power creates. The Supreme Court has the added responsibility of interpreting the intent of the constitution in certain cases.

Ireland's government, like that of the United States, is divided into three powers. In three other ways, however, it is quite different than U.S. political practices. In Ireland, the people elect officials through proportional representation. Political parties have to form coalitions in order for the government to function. The Irish people also vote for members of parliament in the European Union.

In Ireland's proportional representation system, several candidates run for each seat or group of seats. When the Irish vote, each person ranks the candidates first, second, third, and so on. When a vote is counted, if the voter's number one candidate already has enough votes to win, that voter's vote will go to the candidate he or she ranked number two. A proportional representation system has two primary goals. First, it ensures that minority parties are represented in the government. Second, it makes sure that voters' preferences are recognized; so, those votes are not wasted.

There are several political parties in Ireland that have a lot of supporters. After general elections to the Dáil, it is rare for a single political party to hold the majority of the seats in the Oireachtas. Typically, the party that wins the most seats will form a coalition with another party that has similar stances on major issues. These two (or sometimes more) political parties will work together as a coalition government and try to pass legislation. Once the coalition parties find that they cannot work together, the Taoiseach will typically advise the president to dissolve the Dáil and call for new elections.

MEMBERSHIP IN THE EUROPEAN UNION

All countries that are members of the European Union send representatives to the European Parliament in Strasbourg, France. European Parliament elections are held every five years. The European Parliament has different political parties than Ireland. Some parties hold views similar to those of the Irish; others have concerns that are unique to the European Union. Membership in the European Parliament is allocated to each country in the European Union according to population. Ireland currently has 15 of the European Parliament's 626 members. Parliamentary debates are held in 11 official languages, including English, but not the Irish language.

SUMMARY

Politics in Ireland today are marked by an undertone of concern for Northern Ireland and a strong desire among the Irish to bring peace to the island. The civil war in Ireland ended in the 1920s. A guerrilla war, however, continued in Northern Ireland between the Irish Republican Army and the Unionists. At that time, Catholics comprised over 30 percent of the population in the north, but did not have equal rights in government and society.

During the 1960s, a civil rights movement began to sweep the globe. At this time, Catholics in Northern Ireland

sheep scattered throughout the island. As one drives along rural roads in the west of Ireland today, it is common to come upon a flock of sheep in the road. In any given year, there may be more sheep than cattle in Ireland, but cattle remain far more important economically.

Export of animals and animal products is a vital part of Ireland's traditional economy. Its importance continues today, as the country exports about 90 percent of its beef and 75 percent of its dairy production. Ireland also exports sheep and sheep products. A well-known sector of the traditional Irish economy is the production of Irish-knit wool sweaters, especially those made in the Aran Islands, located in the west.

A particularly unique aspect of Ireland's traditional animal industry is today's high-tech stud farms. Ireland is home to numerous stud farms, where many of the world's most successful thoroughbred racehorses are born and bred. Some of the most famous are in County Kildare. Today, this traditional segment of the economy has grown into a multimillion-euro agricultural industry. Irish-born racehorses are running on almost all the world's major racetracks.

While secondary to livestock in land use, Ireland's agricultural economy includes crop production. People perceive Ireland as being a land where potatoes dominate agriculture. While that was true prior to the Great Famine, it is not true today. The major food crop grown today is barley. It leads all crops in acreage planted and is used both as livestock feed and as a major ingredient in many of Ireland's famous beers.

The base of all beers is grain. The grain, whether barley, rice, or something else, is malted, milled, and mashed during the brewing process. One of Ireland's principal products is beer, and Guinness is by far the land's leading drink. It is a key part of Ireland's social life and an icon of Irish culture. The first Guinness brewery is located in Dublin, and every Guinness brewed throughout the world starts from an extract brewed at the Dublin location.

Guinness, famous for its stout (a heavy, dark ale with a strong malt flavor), is headquartered in Dublin. Guinness is a major customer for Ireland's farmers who provide the barley and other grains which brewers malt, mill, mash, and ferment into beer, ale, and stout.

The best lands for crop production are in Ireland's south and southeast (near Dublin). Here can be found field after field of barley. Other major crops important to the economy include wheat, oats, and hay. Ireland also grows sugar beets, turnips, tomatoes, and a variety of seasonal vegetables.

Seasonal vegetables are consumed locally. The Irish make wonderful soups and are great soup lovers. Whatever vegetable is in season on any given day is pureed into vegetable soup. One day vegetable soup could be potato, the next day leek, and the next turnip.

Although agriculture still dominates the Irish landscape, Ireland, like the United States and most of Europe, continues to experience a decline in the number of farms and farmers. As a result, the average farm size is increasing. Many Irish farmers and/or their spouses are employed in non-farm jobs in order to earn sufficient income.

In addition to agriculture, three other sectors of Ireland's economy fall within the traditional economy of Ireland: forestry, fishing, and mining. Each of these sectors is experiencing a renewed focus as part of Ireland's overall economic development plan.

THE FOREST INDUSTRY

The Irish people and their government are making major efforts to achieve reforestation. The goals are to diversify land use, reduce timber imports, restore natural vegetation zones, enhance the economy, and provide a variety of wood and lumber products for the island's newer industries. The successes are a result of intensive efforts at reforestation, not nature. Today's forests are primarily coniferous trees that grow quickly and provide softwoods for timber and paper industries. Most of the forests in Ireland are under government control.

FISHING

Historically, fishing has played a small role in Ireland's economy. Although fishing villages existed along the coasts, the fishing was part of a localized, subsistence economy. Today, Ireland is developing a fishing industry that focuses on two production fronts—the sea and the rivers. From the sea, the fishing industry harvests saltwater fish species such as cod, herring, plaice, and mackerel. The catch from coastal waters includes lobster, crab, mussels, prawn (shrimp), oyster, eel, and crawfish. Fish farming is expanding in selected coastal inlets. On rivers and lakes, fishing is for commerce and for sport. The River Shannon is especially noted for salmon. Sport fishing is popular near Conna in County Cork on the River Bride. Whether at sea or on rivers, in Ireland, most of the commercial fish catch is exported.

MINING AND MINERALS

While mining has not been a major player in Ireland's economy, it provided the minerals needed during the Bronze and Iron ages and early manufacturing. Ireland's landscape is dotted with small quarries. Most quarries provided non-metallic minerals such a limestone, sand, and gravel.

Marble is a metamorphic rock created when limestone experiences great heat and pressure over time. Ireland has over 40 different patterns of marble. The greatest concentration of marble quarries is in the Connemara region. The beautiful green shades of Connemara marble are the most popular colors. Marble from the quarries is shaped into jewelry, architectural stones, and gift products in the city of Moycullen in the Connemara region.

THE HIGH-TECH ECONOMY

The high-tech economy is composed of industries and economic activities that have achieved prominent importance in the Irish economy within the last twenty-five years. Some of the industries have existed on the island for far longer periods of time, but until now, they were not major factors in the economy. Their growth, development, and importance correspond with changes in technology, investments, planning, development, and the economy.

The story of the high-tech economy tells of how Ireland, one of Western Europe's "poor four" in the 1980s, became one of the fastest-growing economies in the world by 2002. Many factors were involved in bringing about this economic miracle. Perhaps none was more important than the government's decision to join the European Union. The European Union has promoted the planning and development of a modern transportation network, has allocated funds for both tourism and industrial development, and has offered economic assistance programs throughout the country.

Foreign corporations that have invested in Ireland, establishing businesses and factories, have also played an important role. At the same time, the Irish government also deserves credit for this success. For over fifty years, it developed plans and programs that it believed would provide a better standard of living for the Irish people. Some were immediately successful, some failed, and some took a long time for results to be seen.

IRELAND TODAY

Today, Ireland's economy is among the strongest in Europe. In recent years, the growth of Ireland's Gross Domestic Product (GDP) has exceeded that of Germany, France, and the United Kingdom. It has even exceeded the growth of the United States.

A lot has changed for Ireland's economy in the last decade. Changes began after the country joined the European Union in 1973. Six countries—France, Germany, Italy, Belgium, the Netherlands, and Luxembourg—created a steel and coal community in the 1950s. The European Union began as a way to trade goods across country borders without taxes and quotas. Over time, the European Union has grown to be an economic and political union among 15 countries, including Ireland and the United Kingdom.

One of the main goals of the European Union has been to help its poorest members—historically Ireland, Portugal, Spain, and Greece—gain wealth. Through its many programs, the wealthy European Union countries help build infrastructure (roads and bridges) and communications (phone and cable lines) in the poorest regions. This, the European Union hoped, would help bring business into the poorest areas of Europe.

Ireland offers a perfect door into the European Union, particularly for American corporations. Most North Americans speak English and American schools have stressed English over foreign languages for decades. In Ireland, American companies have found employees whose primary language is

English. These companies have been attracted by many things, including Ireland's less expensive land and its new infrastructure and communications (built with European Union funds). Today, many migrants come to Ireland in search of jobs. Ironically, many of these immigrants arrive from the United Kingdom.

The European Union has invested billions of euros in the Republic of Ireland. The money has been invested in job creation, upgrading transportation and utilities, developing tourism, and protecting the environment. It has changed the economy of Ireland from one of Europe's poorest countries to one of its most vital. Northern Ireland, however, remains one of the poorest regions in the United Kingdom. It now lags well behind the Republic of Ireland in infrastructure development, tourism, and technologies. The primary reason, of course, is the ongoing civil conflict and related acts of terrorism that continue to occur there.

One of the most important factors contributing to Ireland's recent success is the development of an excellent educational system. For centuries, the Irish people were denied full access to learning. Yet the children of Irish people who moved to other lands did well in school and eventually prospered in a great variety of economic ventures. This helped the Irish people recognize the importance of learning. It also encouraged them to support the development of a quality educational system for the Republic of Ireland. For many years the country did not have enough jobs for its people.

The education system was knowingly educating Irish children so they could emigrate (an emigrant is one who leaves) to other lands and earn a living. Recently, Ireland's new and expanding industries have successfully retained technical school and university graduates at home. Some businesses and industries have even sent recruiters to Boston and New York City to persuade thousands of young Irish to return home to work.

Chris Holm, chairman of Dublin-based IONA Software, represents the tremendous growth of the Irish economy over the last several decades. Ireland is the world's second-largest software producer and also is the European operations, manufacturing, or service headquarters for nearly all of the world's best-known names in computer hardware.

BUSINESS AND INDUSTRY

Today, business, industry, and tourism are the major forces in Ireland's economy. The country is a major European and world manufacturer for a wide variety of high tech and new industrial products. It is the European center for computer hardware and software. Virtually every big player in this industry has a presence in Ireland. The country serves as the headquarters, manufacturing center, or service center for many international companies. They include Microsoft, Dell, IBM, Apple, Hewlett Packard, Motorola, Seagate, and Mitsubishi, to name just a few. Over 1,200 different companies from around the world have selected Ireland as their center for European operations. The greatest concentration of industries is in Dublin and surrounding communities such as Malahide and Swords. Increasingly, however, they are becoming widely scattered across the Irish landscape.

Related industries are now developing. These industries produce equipment, services, or other essentials for firms involved in data processing, office machinery, communications, and electrical products. Other growth categories include cement and related construction and building materials needed for the expansion of housing and highway building. Furniture manufacturing for both office and home is of growing importance. Ireland's pharmaceutical and chemical industries are thriving. So, too, is petroleum refining and natural gas distribution.

TOURISM

In recent years, Ireland has become a major European tourist destination. Tourists from America, Canada, Australia, New Zealand, the European mainland, the Middle East, and Japan, in particular, increase in number annually. Such increases have stimulated the building of transportation facilities, hotels, and other accommodations to which tourists are accustomed.

For the last quarter century, Ireland has vigorously promoted tourism. Their effort has paid off, because tourism is now the largest single contributor to the country's new economy. More than 6 million people visit Ireland a year, and the number continues to grow. Tourism, in the form of tourist expenditures and related taxes, generates over 6 billion euros a year for the Irish economy. Tourism employs one out of every eight Irish workers. It also employs a significant number of temporary summer workers from other countries. One significant benefit of tourism is that it often attracts visitors to small towns, rural areas, and unique landscapes.

Ireland's most important tourist attraction is its people. The Irish people make tourists from around the world feel welcome. Ireland provides the visitor with a land of great beauty, a safe and clean environment, and a "no problem" pace of life. It also offers a variety of recreational and sporting opportunities, especially golf and fishing.

Among Ireland's major tourist attractions are uniquely beautiful natural areas and landscapes such as the Cliffs of Moher and the Ring of Kerry. Cultural attractions include Muckross House, Kilkenny Castle, and the Cobh Heritage Centre. Religious attractions include St. Brigid's Cathedral, Glendalough Monastery, and the Shrine at Knock. Other attractions include handicraft centers, museums, historic sites, architectural achievements, tidy towns, and traditional homes and communities.

The rise in tourism has affected another area of Ireland's economy, the production of traditional goods. Tourists want souvenirs, and the industries that produce Irish-knit wool sweaters, linens, and crystal are expanding to meet the tourist demands.

For example, the famous Aran wool sweaters are hand-woven on the Aran Islands. They have become increasingly popular with tourists and are now sold throughout the world. As a result of the increased demand, similar wool sweaters currently come from cottage industries located almost any-where in Ireland. Today, Ireland dresses people from head to toe. It is noted for fine woolen suits and blazers, scarves, and hats. It produces cotton slacks and dresses and cotton and nylon hosiery. It also produces several lines of footwear.

Irish lace and linens are also cooperative cottage industries centered at Carrickmacross. The prices paid for lace and linen products now reflect the time and artistic talents required to produce them.

Another traditional Irish product in great demand is beauti-ful pottery and china produced in several communities. Perhaps the most valuable traditional product today is crystalware. This industry centers at Waterford where it began in 1783. In the mid-1900s, the crystal industry fell upon hard times and almost

Waterford Crystal is the most famous of Ireland's crystalware producers that trace their origins back to the late 1700s. Beautiful Irish pottery and china, along with crystalware produced in Waterford, Tipperary, and Galway, are popular in the world's advanced economies.

ceased. Waterford Crystal rebounded, however, and has gained spectacular acceptance on the world market. Today, other crystal manufacturers have successfully emerged in Tipperary, Galway, and elsewhere to meet the demands of high-tech economies.

BANKING AND FINANCE

The European Union, industrialization, and tourism expansion have stimulated the growth of Ireland's banking and financial services. The Central Bank of Ireland is located in Dublin and the basic unit of currency is the euro. The Central Bank of Ireland is also a member of the European Central Bank (ECB). Ireland swiftly and almost seamlessly converted from the Irish pound, or punt, to the euro in 2002. Commercial Irish banks have developed an extensive network to serve the needs of the people. They also have established operations in a variety of other countries around the world. At the same time, major foreign banks have established operations in Ireland, primarily Dublin.

WORLD TRADE

Ireland's major trading partners are other European Union countries, especially the United Kingdom, Germany, and France. The United States and Japan are also major trade partners. In addition to food and beverage products, major exports include computers and electronic equipment, pharmaceuticals, chemicals, and clothing. Principal imports are machinery, trucks and automobiles, iron and steel, petroleum products, and food products.

TRANSPORTATION

The transportation infrastructure of Ireland is undergoing significant change. It has 54,476 miles (87,674 kilometers) of highway. Almost 95 percent of the road system is paved. Many of the roads are very narrow, providing access to isolated pastures, farmsteads, and communities. Today, the highways connecting the major cities and towns are undergoing substantial upgrading to handle the increased traffic resulting from the expansion of trucking, automobiles, and tour coaches. Until recently, all roads passed through the city center of each community. The new highways include bypasses, which direct traffic—and some say business—away from the traditional economic center.

Ireland has two types of railroad, broad gauge and narrow gauge. Both differ from the standard gauge found in the United States; one is larger and the other narrower. The railroads are government-owned and operated. The broad gauge railroad is administered by the Irish Transport Company. It links major cities and ports through approximately 1,200 miles

(1,931 kilometers) of track. The narrow gauge track is operated by the Irish Peat Board. It is used to transport peat to briquette processors and power plants. Narrow gauge rails total 850 miles (1,368 kilometers) of track.

The railroads and highways lead to Ireland's major port facilities. These port cities serve as the land's export and import centers. They are also harbors for the ferryboats that transport motor vehicles between Ireland and other European gateways. The major port cities are: Dublin, Cork, Galway, Limerick, Waterford, Drogheda, Dun Laoghaire, Arklow, Foynes, and New Ross. About 75 percent of Ireland's world trade flows through these seaports.

Ireland has developed major international airports at Dublin, Shannon, Cork, and Knock. Dublin and Shannon are by far the two most important air hubs in Ireland, with numerous daily flights to major world cities. Ireland also has 40 small paved and unpaved airport and landing facilities. Well in excess of ten million passengers a year fly through Ireland's airports.

Pipelines for natural gas transmission and distribution are becoming increasingly important in Ireland. Natural gas and petroleum must move from the coastal fields and refineries to distribution centers. The island is experiencing increasing demands for natural gas and propane for home and industry.

TELECOMMUNICATIONS INFRASTRUCTURE

One of the most important factors in the recent economic growth of Ireland is its telecommunications infrastructure. The development of a "world class" digital telecommunications system put Ireland at the forefront of European Union expansions. The Irish government operates or authorizes all communications services. This includes the postal service, telephone, telegraph, radio, television, Internet service providers, optical fiber, and telecommunications. Deregulation now allows telecommunication companies to compete for Ireland's business. This is expected to further expand such facilities.

Government agencies, or statutory bodies, operate all Irish postal, telegraph, telephone, radio, and television services.

SUMMARY

The story of Ireland's traditional and high-tech economies is one of spectacular change. In just 25 years, Ireland has escaped thousands of years of economic suppression and poverty to become one of the world's fastest-growing economies. In a short time, it has moved from an economy based primarily on agriculture to a diversified high-tech economy that includes manufacturing, tourism, transportation, banking, and trade in its economic story. The next chapter of this story will address how Ireland can continue to improve its infrastructure, further expand educational programming, protect the environment, raise the standard of living, adapt to cultural change, and enhance its economy.

Although there is an undercurrent of civil strife known as "The Troubles," day-to-day life in Ireland is best characterized by a "no problem" outlook. The streets in Dublin are filled with people who reflect Irish friendliness, pride in Irish culture, and the energy that fuels Ireland's fast-growing economy.

7

Living in Ireland Today

Fadó, fadó, the Celts began the tradition of storytelling in Ireland. Storytelling today is just one aspect of Irish culture and daily life that is celebrated in annual festivals throughout the country. Irish festivals can last one or two days, or even a month. Among the most notable festivals are the storytelling festivals in the west, the Galway arts festival, the Kilkenny arts week, and the Liisdoonvarna matchmaking festival.

Wandering through one of the country's many festivals, a tourist would frequently hear the phrase "no problem." In Ireland, tourists will hear the phrase uttered by hotel clerks, waiters, and people on the street. This phrase and the many festivals create a perception that life in Ireland is slower paced, that the Irish have a lot to celebrate, and they have few problems.

THE TROUBLES

Ironically, the undercurrent of life in Ireland is a painful problem that the Irish have described as "The Troubles." The Troubles are indeed the most important factor affecting daily life in Ireland and Northern Ireland. Their roots can be traced back to the efforts of King Henry VIII, William of Orange, and Cromwell to destroy the religious freedom of the Irish. Over time, these roots branched out to include economic, social, intellectual, and cultural influences as well. Unfortunately, these destructive influences did not die with Irish independence. They continue to be woven through every fabric of Irish life even today. While the effects and events of The Troubles are far greater in Northern Ireland, they still have very real impacts on daily life in Ireland. Perhaps the best way to understand this is to compare it to a family in which two members are almost constantly fighting. The conflict continuously affects all other family members. As long as the fighting continues, it is a painful problem to all family members no matter where they live.

The island of Ireland is home to two distinct governmental units: the independent Republic of Ireland and Northern Ireland, which is part of the United Kingdom. The majority of people in the North are Protestant Unionists (or loyalists). Historically, they have used a variety of political actions to maintain control of the land and to restrict the civil rights of the North's Catholics. In the 1960s, a call for civil rights for the minority Catholic population developed in Northern Ireland. It called for an end to anti-Catholic discrimination. It was bitterly opposed by the Unionists. The result was troubling disorder and rioting.

During The Troubles, more than 3,000 people were killed. Most fatalities were of civilians on both sides. The deaths included members of extremist groups and British soldiers. Thousands of people also were injured. Homes were bombed and burned. Families were destroyed. Hatreds intensified.

Irish students enter school at five years of age. At seven, they enter the first class and stay through the sixth class before entering the second level which takes five or six years to complete. A successful score on the final exam at the completion of secondary schooling opens the way to technical school or university.

HEALTH AND SAFETY

Health and safety are also important aspects of Irish life. Ireland has a national health care and insurance system that, depending on income, provides a full range of services that vary from free to full payment. Safety is the responsibility of the police. Overall crime in Ireland is low. There are exceptions in some urban zones where drugs are a problem.

GENEALOGY

Stemming from the Great Migration and Ireland's huge tourism industry, a growing sector of Ireland's culture has an interest in genealogy. Among the Irish people, family history is passed down orally from generation to generation. For

Americans, Australians, and others, often all they know is that a grandfather or great-grandmother was born in Ireland. They are the people who want to know more and are willing to pay to find the answers. In the process, they also often find relatives in Ireland that they did not know existed.

Genealogy has provided well-paying careers to some professional researchers in Ireland. If people do not want to pay researchers to unlock the history of their families in Ireland, the townland, a unique geographical aspect of genealogy in Ireland, may be the key. Ireland is divided into over 60,000 pieces of land called townlands. They range in size from about 3 acres to 600 acres. If someone seeks genealogical information about relatives in Ireland and knows the townland of his ancestor, it is much easier to find records. Also, if one knows the townland, it is possible to visit the exact place in Ireland where one's ancestors lived and died.

DAILY LIFE IN IRELAND

Daily life in Ireland begins around 8 A.M. The Irish stay up late at night and get up later in the morning than do most Americans. By about 9 A.M., people are on the streets. The traffic rush hits the roads about the same time. Mothers with prams (baby carriages) are walking their older children to school or going to the grocery. School children of all ages are on buses or walking to school. Students wear sweaters and pants or skirts displaying the colors of their school. In the country, farmers gather along the roads with milk pails awaiting the arrival of dairy trucks. By 10 A.M., everyone is at school, work, home, or shopping. At noon, the Irish have lunch. It often consists of a bowl of soup and a sandwich. School children leave the school grounds and some eat lunch in restaurants, cafeterias, or pubs. After lunch, the Irish return to work, school, or shopping.

The school day ends in the late afternoon, and the workday between 5 and 6 P.M. There is another brief traffic rush hour in the city centers and on the roads. About 8 P.M., or later, the Irish enjoy their evening dinner. Watching television is not an important evening social activity of the Irish. They only have a few channels (one in Irish only) and the shows are generally not comparable to nightly television in the United States.

Nightlife for adults in Ireland begins about 10 P.M. The center of this activity is the pub. It is the social center of the neighborhood or community. It is the place for Guinness, friendly people, storytelling, music, *craic* (pronounced "crack," this is Irish for "fun"). Live music in an Irish pub seldom begins before 10 P.M. By midnight, the people are on their way home, where they then sit around the table or fireplace and discuss the events of the day before bed.

SUMMARY

As the story of daily life in Ireland concludes, one can see that life in Ireland is in some ways very much like living in America and that Irish life contains some unique and wonderful exceptions that are truly Irish. They include a great range of cultural and social impacts such as The Troubles, the intellectual stimulation of their schools, legends and stories, and the physical bawdiness of Irish sports. The Irish people enjoy life, their pubs, and their festivals. More importantly, they enjoy each other. They have built a society and a lifestyle that makes Ireland uniquely Irish.

In Dublin, children join a rally for peace. All people in the Republic of Ireland as well as Northern Ireland hope for peace. In the late 1990s, several compromises reduced the violence. These steps minimized support for terrorism, but have not completely resolved the conflict.

8

The Future of Ireland

Each chapter of this geography of Ireland has told a story about Ireland's physical, cultural, historical, economic, and political, geographies. Knowing more of Ireland through these stories may have changed some perceptions of the country while strengthening others. Visitors to Ireland would experience many of the stories told in this book and would also form new perceptions of the country.

THE PHYSICAL GEOGRAPHY

The story of Ireland's geography began with an analysis of the island's physical geography. The island of Ireland is shaped like a cereal bowl, low and level in the middle and surrounded by higher hills and low mountains on the rim. For its size, the terrain of the island is fairly complex. The higher elevations were formed by two major European mountain-building forces. The bottom of the bowl or central lowlands is a limestone area built up from the ocean floor.

Virtually the entire island was glaciated, leaving glacial drift and glacial landforms such as drumlins.

IRISH CULTURAL HISTORY

The story of Ireland's geography became more complicated when people arrived. Once people interact with the physical environment, cultural landscapes begin to develop. The first known human occupants arrived in Ireland about 10,000 years ago. At that time, the physical environment of the island was changing from a post-glacial landscape to oak forests. About 8,000 years ago, the first Neolithic peoples arrived. They cleared parts of the forests to farm. They also built the fascinating passage tombs of the Valley of the River Boyne. Around 4,000 years ago, Bronze Age people from Mediterranean Europe arrived in Ireland. They brought knowledge of metal making, jewelry, as well as new tools and weapons to Ireland.

The Celtic people and the Iron Age reached Ireland somewhere between 2000 and 1000 B.C. Their arrival changed Ireland forever. With the Celts came iron tools and weapons, kingdoms, power, and wealth. The Celts controlled the land until 432 A.D., when Christianity, the most powerful force in Irish history, arrived in the person of Saint Patrick. Christianity again changed Ireland. The Irish became Christians. Church leaders were more powerful than were kings and chieftains. Monasteries dotted the Irish landscape and were among the major European centers of learning.

The arrival of the Vikings in the ninth century brought new challenges to Ireland. The invaders attacked farms and monasteries seeking food and wealth. They founded forts that became Dublin, Galway, and other cities. They intermarried with the Irish and reintroduced weaponry, shipbuilding, and warfare.

In 1014 A.D., Brian Boru and his armies chased the Vikings from the land. Unfortunately, freedom was short-lived. About 150 years later, armies of England's King Henry II claimed Ireland for his empire. This was the beginning of 800 years of English occupation.

During this time, smaller and somewhat weaker abbeys replaced powerful monasteries. Eventually, the English even tried to forcefully replace the Catholic Church with the Church of Ireland. Over the years, the British occupiers denied the Irish people civil and economic rights. Through penal laws and through land redistribution, the British tried to destroy Irish religion, culture, society, and traditions.

In the late 1800s and early 1900s, cries for Irish independence began rising from the land. In 1916, on Easter Sunday, 1,500 Irish rebels raised an Irish flag, declared independence, and occupied key buildings in Dublin. The British countered, killing and arresting many of the rebels. This was followed by a series of conflicts that were terminated on July 11, 1921, with the creation of the Irish Free State and the United Kingdom of Great Britain and Northern Ireland. Unfortunately, while this was probably the best possible solution for the time, it also became the basis for a continuing series of conflicts. In the new Irish Free State, partisan power conflicts lasted until the establishment of the constitution in 1937 and the Republic of Ireland in 1949.

Conflicts continue today in Ireland. These conflicts occur primarily in Northern Ireland, but they can happen anywhere on the island. They are called The Troubles. The Troubles are the result of an ongoing civil rights and economic power struggle among Unionists, Nationalists, and Republicans. The parties have entered a tentative accord, but The Troubles are still an undercurrent in the daily lives of the Irish.

The story of Ireland and the Irish people is indeed exciting. It shows the development of the people, their culture, and society. For hundreds of years, the Irish have fought for freedom. They have consistently shown a strong dedication to the preservation of their religion, culture, and independence. Many of the cultural traits associated with Ireland date back centuries. During the independence movement of the late 1800s and early 1900s, the Irish worked to preserve and reinvigorate their language, religion, music, dance, writing, art, architecture, and past times.

Up to now, this story has focused on the reciprocal relationship between the physical and cultural landscapes of Ireland. Within this context, the Irish live their daily lives and create their own unique cultural environment. The cultural environment of Ireland today shows a country whose population is changing. Its government is interacting at new levels, its economy is rapidly growing, and the people have developed a way of life that copes with all of these changes.

IRISH POPULATION AND GOVERNMENT

The population of Ireland is changing, having reached 3.9 million in 2002. This figure represents the highest population count since the Great Famine of 1846 and the heavy out-migration that followed that tragic event. Population growth in Ireland is expanding at a rapid rate. Almost half of the growth, however, is the result of in-migration from foreign lands. The greatest population growth is being experienced in the counties that surround Dublin as the urban sprawl of Ireland's capital expands outward. Cork remains the second largest city, but Galway is challenging it for this position.

The government of Ireland is working not only within the country; it is now interacting on a new dimension, the European Union. Ireland, like America, has three powers of government: the executive, legislative and judicial. Ireland has both a president and a Taoiseach (prime minister). The presidency is primarily a ceremonial position within the legislative power. The day-to-day operation of the government is under the leadership of the Taoiseach, in the executive power. Ireland's legislative body is the Oireachtas. It has two houses, the Seanad Éireann, or Senate, and the Dáil Éireann, or House of Representatives.

Ireland has benefited immensely from membership in the European Union, as measured by its growing economy, new roads and communications technologies, and its newly preserved cultural artifacts. The people of Ireland strongly support their membership in the European Union and recognize

At a European Union (EU) summit meeting in Seville, Spain, on June 21, 2002, heads of state and foreign ministers, including Irish Prime Minister Bertie Ahern, stand for a group photograph. Membership in the EU has changed Ireland's economy. It went from being one of Europe's "poor four" in the 1980s to an economy whose growth rate now exceeds that of the United States, the United Kingdom, Germany, and France.

that they have benefited since becoming members in 1973.

As a member of the European Union, the Irish government works within the country, within the counties, and also within the context of European politics. The Irish government is an active, involved, and respected member of the world community. Ireland participates fully in international organizations, agencies, and efforts. Ireland's government, churches, and people are particularly noted worldwide for service to the poor, the sick, and the hurting.

IRISH ECONOMIC DEVELOPMENT

Ireland's economy is growing rapidly. Ireland has made tremendous economic changes in a relatively short period. For centuries, Ireland survived by farming the land and processing its agricultural resources. Today, as in the past, Ireland is agriculturally a land of livestock, but with some key crops. Beef and dairy cattle comprise the most important part of the agricultural economy. Historically, a New World (the Americas) crop, the potato, became the key food crop of the Irish people. When the potato blight and resulting famine hit in the 1840s, Ireland suffered tremendous hardship, loss of life, and out-migration. Today, the key crops in Ireland are barley, wheat, and flax.

Through efforts of the Irish government, membership in the European Union, and investment by foreign corporations, manu-facturing has increased dramatically in the last 25 years. Today, Ireland is a key player in the world's high-tech computer hardware and software industries. It is a center for pharmaceuticals, chemi-cals, and communications equipment. It is also experiencing the massive development of important cement and construction industries that are building the new infrastructure. Ireland is also a land of internationally valued, traditional products such as Waterford Crystal, Aran sweaters, and Guinness stout.

Perhaps the most rapidly changing aspect of Ireland's economy is the transportation infrastructure. Fueled by funds from the European Union, wide modern high-speed highways are replacing the old, narrow roads. Airport and port facilities are constantly being upgraded. This new infrastructure not only encourages industrial development, it also facilitates spectacular tourism expansion as well. All of these developments have also increased the importance of banking and finance in the Irish economy. Ireland moved from a land of high unemployment to one experiencing labor shortages in less than 25 years. Today, the country is a rising economic and manufacturing force in Europe's economy.

IRISH LIFE

The final topic of this adventure is the story of life in con-temporary Ireland. Life in Ireland is both complex and relaxed.

This combination is a result of the sorrowful impact of the ongoing Troubles, cultural practices, and daily life that point to a strong faith and a "no problem" attitude.

Ireland has been blessed with a strong oral tradition that preserved its myths, legends, and genealogy. It is a land of outstanding literature, poetry, theatre, music, and dance. It is also a land that through architecture and artifacts has made major efforts to preserve the treasures of its past. Ireland is a land of sport and celebrations, with a fanatical love of its native games of Gaelic football and Irish hurling, as well as a national commitment to feasts and festivals. Irish culture provides a welcoming environment to tourists who are most often people of Irish descent seeking to learn more of the land and peoples of their ancestry. Irish and tourists alike enjoy the "crack" (fun), music, and social benefits of pub life.

THE FUTURE OF IRELAND

The chapters yet to be written in the story of the geography of Ireland will surely address many of the changes the Irish economy has experienced in the last 25 years. Storytellers of the future will certainly have many characters and plots to discuss. Certainly they will incorporate the reinvigorated Irish culture of the 1900s, the Irish education system, and entry into the European Union. They also will tell of The Troubles, the tourists, and the new Irish economy.

The outcome of the story is uncertain. Surely, they will tell of Ireland's changing physical and cultural landscapes. No doubt they will include reference to the way the Irish people identify themselves, the way the government gets involved in European and world politics, and the way the Irish maintain and change their culture and daily lives. Having learned the many stories of Ireland, it is clear that when they are faced with the challenges of tomorrow, the Irish people will assuredly say "no problem!"

Facts at a Glance

Name	Republic of Ireland.
Background	In 1921, Ireland gained independence from the United Kingdom. Northern Ireland remained part of the United Kingdom of Great Britain and Northern Ireland. In 1973, it joined the European Community (today's European Union). In 1998, it approved a peace accord on Northern Ireland.
Relative Location	Off the west coast of Europe, surrounded by Atlantic Ocean and Irish Sea.
Absolute Location	53 degrees north latitude, 8 degrees west longitude.
Area	27,136 square miles (70,282 square kilometers), slightly larger than West Virginia.
Coastline	3,500 miles (5,633 kilometers)
Climate	Marine west coast. Mild, moist climate lacking temperature extremes.
Terrain	Central lowlands surrounded by hills and low mountains.
Elevations	Lowest Point, sea level; Highest Point, Carrauntuohil 3,414 feet (1,041 meters).
Minerals	Limestone, peat, lead, zinc, natural gas, and copper.
Population	3,917,336 (2002).
Religions	Roman Catholic, 91.6% (1998). Church of Ireland, 2.5% (1998).
Languages	Irish is national language and first language. English is second language and is the primary language by use.
Literacy	98%.
Capital	Dublin.
Counties	Twenty-six counties: Carlow, Cavan, Clare, Cork, Donegal, Dublin, Galway, Kerry, Kildare, Kilkenny, Laois, Leitrim, Limerick, Longford, Louth, Mayo, Meath, Monaghan, Offaly, Roscommon, Sligo, Tipperary, Waterford, Westmeath, Wexford, and Wicklow.
Independence Day	December 6, 1921.
National Holiday	Saint Patrick's Day, March 17.
Flag	Three equal vertical bands of green (hoist side), white (center),and orange.

Industries	Computer hardware and software, brewing, textiles and clothing, chemicals, pharmaceuticals, machinery, transportation equipment, glass and crystal.
Exports	Machinery, computers, chemicals, pharmaceuticals, and livestock.
Imports	Data processing equipment, machinery, chemicals, petroleum, textiles and clothing.
Currency	Euro.
Railways	Two types of track. Irish Transport Company uses broad gauge track and links cities and ports with about 1,200 miles (1,931 kilometers) of track. Irish Peat Board uses narrow gauge track to transport peat to briquette processors and power plants. Narrow gauge track totals 850 miles (1368 kilometers) of track.
Highways	Total of 54,476 miles (87,674 kilometers). 95 % of road system is paved.
Waterways	Two canal systems totaling 270 miles (435 kilometers) long.
Airports	Major international airports at Dublin, Shannon, Cork, and Knock. Dublin and Shannon are the two most important air hubs. Also has and additional 40 small paved and unpaved airport and landing facilities.
Major Ports	Dublin, Cork, Galway, Limerick, Waterford, Drogheda, Dun Laoghaire, Arklow, Foynes, and New Ross. About 75% of Ireland's world trade flows through these seaports.

History at a Glance

12,000 B.C.	Ice Age ends. Island begins process of environmental change from arctic to temperate climate.
8000	First known human occupants arrive on the island of Ireland.
5000	Ireland covered by oak and elm forests.
4000	Neolithic people reach and begin occupying Ireland.
2000	Bronze Age people arrive in Ireland.
1000	Celtic people arrive and settle.
500	Celtic kings and chieftains battle for position of high king.
430 A.D.	Palladius is sent to Ireland by Pope Celestine as its first missionary.
432	Saint Patrick brings Christianity to Ireland and establishes the faith.
550	Growth of Celtic monasteries.
795	Viking invasions of Ireland begin.
841	Vikings build fort that becomes Dublin.
1041	Brian Boru defeats the Vikings and restores Irish rule.
1170	Anglo-Normans under Strongbow capture Wexford and claim Ireland for King Henry II of England.
1348	Black Death kills off one-third of the Irish population.
1536	King Henry VIII, having broken from the Roman Catholic Church, declares himself head of the Church of England and establishes Church of Ireland.
1582	Ireland is divided into 32 counties.
1592	Trinity College established.
1649	Cromwell arrives, attacks, destroys, and kills thousands of Irish people.
1690	William of Orange defeats King James II at Battle of the Boyne.
1800	Ireland through Act of Union becomes part of the United Kingdom of Great Britain and Ireland.
1828	Catholic and Presbyterian Emancipation Act passed under leadership of Daniel O'Connell.
1845	The Great Irish "Potato" Famine begins. Great out-migration, primarily to North America, begins.
1877	Home Rule Party and other nationalist movements gain followers.
1884	Gaelic Athletic Association established.
1904	Abbey Theater opens.

1905 Sinn Féin (we ourselves) founded by Arthur Griffith.

1916 Easter Rebellion.

1920 Bloody Sunday (Sunday, November 21, 1920).

1921 Anglo-Irish treaty partitions Ireland into Irish Free State and Northern Ireland Administrative District.

1922 Irish Civil War.

1926 Eamon de Valera establishes Fianna Fáil (soldiers of destiny) political party.

1932 De Valera elected Taioseach (Prime Minister).

1937 New Constitution ratified.

1939 Ireland neutral during World War II.

1949 Republic of Ireland established and recognized by United Kingdom.

1955 Ireland joins United Nations.

1967 Northern Ireland Civil Rights Association organized.

1972 British soldiers killing of 13 people in North memorialized as Bloody Sunday.

1973 Ireland joins the European Economic Community (later the European Union).

1979 Pope John Paul II visits Ireland.

1985 Anglo-Irish Agreement signed by UK and Ireland.

1991 Mary Robinson becomes Republic of Ireland's first female President.

1994 Irish Republican Army and Protestants enact cease-fires.

1998 Good Friday Agreement signed.

Further Reading

Carter, R.W.G. and A.J. Parker, eds. *Ireland: A Contemporary Geographical Perspective.* New York: Routledge, 1989.

Central Intelligence Agency, *The World Factbook 2001.* Washington D.C.: CIA, 2002
http://www.cia.gov/cia/publications/factbook/

Central Statistics Office. *Census 2002 Preliminary Report.* Dublin, Ireland, 2002.

Department of Foreign Affairs. *Geography of Ireland,* Dublin, Government of Ireland, 2002
http://www.irlgov.ie

Irish Internet Hub
http://larkspirit.com/general/irishhub.html#culture

Gerard-Sharp, Lisa, and Perry, Tim. *Ireland: Eyewitness Travel Guides.* New York: Dorling Kindersley Publishing, Inc., 1995.

MacLaughlin, Jim. "Nation-building, social closure, and anti-Traveller racism in Ireland." *Sociology,* 33 (1999): 129.

Marketing Services Division, *Ireland Vital Statistics,* Dublin, IDA Ireland, April 2002
http://www.idaireland.com/news/pdf/vitalstatsmay02.pdf

Nolan, Mary Lee. "Irish Pilgrimage: The different tradition." *Annals of the Association of American Geographers,* 73 (1983): 421-438.

Orme, A.R. *Ireland.* Chicago: Aldine Publishing Company, 1970.

White, Terence De Vere. *Ireland.* New York: Walker Company, 1968.

Index

Index

page:

8:   New Millennium Images
13: 21st Century Publishing
16: New Millennium Images
21: 21st Century Publishing
25: New Millennium Images
29: New Millennium Images
32: New Millennium Images
38: Hulton/Archive by Getty Images
43: Hulton/Archive by Getty Images
45: Hulton/Archive by Getty Images
48: New Millennium Images
54: New Millennium Images
57: New Millennium Images

Cover: New Millennium Images

59: KRT/NMI
62: Hulton Getty Photo Archive/NMI
67: KRT/NMI
70: John Cogill/AP/Wide World Photos
74: New Millennium Images
78: New Millennium Images
83: John Cogill/AP/Wide World Photos
86: New Millennium Images
90: New Millennium Images
93: Peter Morrison/AP/Wide World Photos
97: KRT/NMI
100: AFP/NMI
105: Yves Logghe/AP/Wide World Photos

About the Authors

EDWARD PATRICK HOGAN is Professor of Geography at South Dakota State University and the State Geographer of South Dakota. He and his wife Joan have visited Ireland six times in the past 7 years. They have visited family and friends, led tour groups, and have witnessed the phenomenal changes that have occurred and continue in Irish society. In addition to his career teaching geography, Ed is the Associate Vice President for Academic Affairs and the Chief Information Technology Officer for South Dakota State University. Ed and his daughter Erin Hogan Fouberg also co-authored *The Geography of South Dakota*, published in 2001 by the Center for Western Studies. Ed has authored numerous articles, television series, and publications related to the types of housing around the world, migration, economic development, and other areas of geographic research. He received the Distinguished Teaching Award from the National Council for Geographic Education and in 1992 was included in the book *Leaders in American Geography*, as one of the 79 people who have most influenced geographic education in the United States. He especially enjoys being with his family, listening to traditional Irish music, and creating works of art.

ERIN HOGAN FOUBERG is Assistant Professor of Geography at Mary Washington College in Fredericksburg, Virginia. She co-edited the book *The Tribes and the States: Geographies of Intergovernmental Interaction* (Rowman and Littlefield, 2002). She is the author of *Tribal Territory, Sovereignty, and Governance: A Study of the Cheyenne River and Lake Traverse Indian Reservations* (Garland, 2000) and co-authored with her father *The Geography of South Dakota*, revised edition (Center for Western Studies, 1998). Aside from her work on jurisdiction in Indian country, Dr. Fouberg has published on the use of writing in geographic education, and has co-edited a special issue of the *Journal of Geography* on education theory in geography. The students of Mary Washington College honored Dr. Fouberg in 2001 with the Mary Pinschmidt Award, for being the professor most likely to be remembered for having an impact on their lives.

CHARLES F. ("FRITZ") GRITZNER is Distinguished Professor of Geography at South Dakota University in Brookings. He is now in his fifth decade of college teaching and research. During his career, he has taught more than 60 different courses, spanning the fields of physical, cultural, and regional geography. In addition to his teaching, he enjoys writing, working with teachers, and sharing his love for geography with students. As consulting editor for the MODERN WORLD NATIONS series, he has a wonderful opportunity to combine each of these "hobbies." Fritz has served as both president and executive director of the National Council for Geographic Education and has received the Council's highest honor, the George J. Miller Award for Distinguished Service.